Geologists' Ass

GEOLOGY OF THE YORKSHIRE COAST

Peter F. Rawson and John K. Wright

with contributions by:
Rodger Connell
Ian Heppenstall
Steve Livera
Ian Starmer
Paul Wood

Guide Series Editor: Susan B. Marriott

4th edition
© THE GEOLOGISTS' ASSOCIATION 2018

CONTENTS

	Page
PREFACE	ii
ACKNOWLEDGEMENTS	ii
LIST OF FIGURES	iv
LIST OF TABLES	v

1 GEOLOGY OF THE YORKSHIRE COAST
1.1 Introduction 1
1.2 The structural framework 4
1.3 Stratigraphy, palaeogeography and environments 7
1.4 The industrial story 20

2 SEISMIC PROFILES 22
2.1 Map of seismic coverage in NE Yorkshire 23
2.2 Regional seismic profile Whitby to Barmston 24
2.3 The Peak Trough 26
2.4 Robin Hood's Bay Dome 28
2.5 The Peak Trough and Cayton Bay faults 29
2.6 Flamborough Head Fault Zone 30

3 EXCURSIONS
Itinerary 1: Staithes to Port Mulgrave 31
Itinerary 2: Runswick Bay 40
Itinerary 3: Saltwick Bay to Whitby 44
Itinerary 4: Robin Hood's Bay and Ravenscar 52
Itinerary 5: Cloughton Wyke to the Hundales 67
Itinerary 6: Burniston and Scalby bays 75
Itinerary 7: Egton Bridge to Grosmont 82
Itinerary 8: South Bay, Scarborough and Cornelian Bay 87
Itinerary 9: Cayton Bay, Yons Nab and Gristhorpe Bay 96
Itinerary 10: Betton Farm, East Ayton and Filey Brigg 107
Itinerary 11: Reighton Gap to Speeton Cliffs 118
Itinerary 12: Thornwick Bay and North Landing, Flamborough 126
Itinerary 13: Flamborough Head 132
Itinerary 14: South Landing to Sewerby 138
Itinerary 15: Langtoft, Foxholes and Staxton Hill 148
Itinerary 16: North Holderness: Barmston south to Mappleton 151
Itinerary 17: South Holderness: Withernsea and Dimlington 156

4 REFERENCES 160

INDEX 171

i

PREFACE

In the 1960s the Geologists' Association produced two field guides that together embraced the Yorkshire coast and some adjacent inland areas. One, covering itineraries from Staithes in the north southward to Flamborough and Sewerby, has remained in print since the initial version by Hemingway *et al.* (1963, revised 1968) followed by revised editions by Rawson and Wright (1992, 2000). The other formed a guide to Hull and surrounding areas (Bisat *et al.*, 1962) and eventually went out of print. That guide included important exposures of Quaternary tills and associated sediments along the Holderness coast. This new edition of the Yorkshire Coast guide has been expanded to include the more important Holderness sections as well as adding a new Lower Jurassic itinerary, Runswick Bay. It thus embraces almost all of the Yorkshire North Sea coastline, together with some adjacent inland exposures.

In this guide we generally use the internationally agreed standard stages, the base of one of which, the Pliensbachian, is formally defined at a horizon in Robin Hood's Bay (see Itinerary 4). There is a problem across the Jurassic/Cretaceous boundary, where the boundary between Tithonian (Jurassic) and Berriasian (Cretaceous) stages is difficult to recognise in northern (Boreal) basins, including the North Sea Basin. Here the regional stages Volgian and Ryazanian are commonly used. The boundary between the two is generally regarded as correlating with a level in the Berriasian stage (e.g. Ogg & Hinnov, 2012) and that is the position adopted here. However, Kelly *et al.* (2015) have suggested that the Volgian stage is simply the Boreal equivalent of the Tithonian.

There is also a difference in the use of fossil zones between Jurassic and Cretaceous workers, reflected in this guide. Jurassic ammonite zones are currently treated by British workers as chronozones (i.e. time zones) while Cretaceous specialists still regard ammonite and other fossil zones as biozones (the boundaries of which may cross timelines). In this text, all chronozones (i.e. the Jurassic ammonite zones) appear in Roman typeface, while the Cretaceous biozones (based on ammonites and other fossil groups) appear in italic typeface.

ACKNOWLEDGEMENTS

In preparing this edition we have once again invited contributions from several other research workers and are very grateful to Rodger Connell (University of Hull), Ian Heppenstall (Hull Geological Society), Steve Livera (formerly Shell International), Ian Starmer (formerly UCL) and Paul Wood (formerly Shell International) for their support. Sadly, Felix Whitham, who made significant contributions on the Chalk in the last two editions, has since died, but his published work forms the basis of revised descriptions expanded and updated by Peter Rawson.

We thank Mark Bateman, Tim Burnhill, Steve Livera and Rory Mortimore for commenting on parts of the text and members of the Rotunda Geology Group (Scarborough) for testing some of the itineraries during the Group's field excursions. Most of the line drawings were prepared by Lynne Blything, formerly of Royal Holloway, and we are very grateful for her tolerance of our numerous modifications! The Geologists' Association kindly provided funding for her work.

We would particularly like to thank Susan Marriott for her encouragement and advice during the preparation of this guide and for all her editorial work while guiding it through the press.

LIST OF FIGURES   Page

1. Outline geological map with itineraries: northern area.   2
2. Outline geological map with itineraries: southern area.   3
3. The structural framework.   5
4. Structural inversion of the Cleveland Basin.   6
5. The palaeogeographical setting.   9
6. Palaeogeography during the Late Devensian (last) glaciation.   19
7. UKOGL seismic coverage along the NE Yorkshire coast.   23
8. Regional seismic profile: Whitby to Barmston.   25
9. The Peak Trough: position of seismic lines shown in Fig. 10.   26
10. Seismic sections across the Peak Trough.   27
11. The Robin Hood's Bay Dome.   28
12. The Peak Trough faults along seismic line TW81-42.   29
13. The Flamborough Fault Zone west of Flamborough Head.   30
14. Map of the Staithes to Rosedale Wyke shore (Itinerary 1).   32
15. Lithic log of the Staithes Sandstone and Cleveland Ironstone formations at
    Staithes.   34
16. The 'Upper Striped Bed' (bed 34 upper) in Jet Wyke.   35
17. The western side of Brackenberry Wyke to Old Nab.   37
18. *Rhizocorallium* in the Main Seam, Brackenberry Wyke.   37
19. Lithic log of the Grey Shale and Mulgrave Shale members: Port Mulgrave to
    Whitby.   39
20. Map of Runswick Bay to Kettle Ness point (Itinerary 2).   41
21. The Millstones and Top Jet Dogger, Topman Steel, Runswick Bay.   42
22. Map of Saltwick Bay to East Cliff, Whitby (Itinerary 3).   45
23. Cliff section at Whitby, as seen from the shore.   46
24. Shale-filled channel in the Saltwick Formation, East Cliff, Whitby.   47
25. East side of Saltwick Nab, Saltwick Bay, Whitby, showing the main marker
    bands.   50
26. Pyritised bivalve (*Pseudomytiloides dubius*) and ammonite (*Harpoceras
    falciferum*), Bituminous Shales, Saltwick Bay.   51
27. Erratic boulder of Shap Granite on the shore in Robin Hood's Bay.   52
28. Map of Robin Hood's Bay and the Peak (Itinerary 4).   53
29. Simplified lithic log of the Redcar Mudstone Formation, Robin Hood's Bay.   55
30. *Diplocraterion* from the Calcareous Shale Member, Robin Hood's Bay.   56
31. Dock excavated in rock platform, Wine Haven, Robin Hood's Bay.   58
32. The Peak Fault from the air.   59
33. Diagrammatic section across the Peak Fault at Ravenscar, as seen from the
    shore.   60
34. Lithic Log of the Mulgrave Shale Member at Peak.   61
35. Lithic log of the Alum Shale Member at Peak.   63
36. Robin Hood's Bay from the Peak.   64
37. Map of localities between Cloughton and Scalby (Itineraries 5 and 6).   68
38. Lithic log of the Ravenscar Group between Cloughton Wyke and Scalby Ness.   69
39. Cliff section in the Gristhorpe Member, Cloughton Wyke.   71
40. The Scarborough Formation, cliffs east side of Cloughton Wyke.   72
41. Diagrammatic section of the lower Long Nab Member, Burniston and Scalby
    bays.   76

		Page
42.	Saurian footprints from Burniston Bay.	77
43.	Rip-up clasts of laminate siltstone in Black's Channel 'E' at Cromer Point.	79
44.	Map of meander belt channels in the Scalby Formation, Scalby Bay.	80
45.	Meander scrolls in meander belt, Scalby Bay.	81
46.	Map of the Egton Bridge, Goathland and Grosmont area (Itinerary 7).	83
47.	Former workings in the Cleveland Dyke, Silhowe.	85
48.	Simplified lithic log of the succession from South Bay to Cayton Bay.	88
49.	Map of localities in the Scarborough to Yons Nab area (Itineraries 8 and 9).	89
50.	Rotunda – the William Smith Museum of Geology, Scarborough.	90
51.	Sedimentary structures in the Moor Grit Member, near the Spa, South Bay.	91
52.	The Moor Grit and Long Nab members, South Bay.	92
53.	View looking SE along Cornelian Bay to Osgodby Nab.	94
54.	Geological map of the strata exposed in the rock platform in Cornelian and Cayton bays.	97
55.	Herringbone cross-stratification in the Gristhorpe Member at Osgodby Point.	98
56.	The Millepore Bed at Osgodby Point.	99
57.	Schematic cross section to demonstrate the development of the Tenants' Cliff block-slide.	100
58.	Cliff section at the south end of Cayton Bay.	102
59.	Red Cliff and the Red Cliff Fault, Cayton Bay.	102
60.	The Helwath Beck Member (Scarborough Formation) at Yons Nab.	104
61.	Schematic cross section of the Corallian rocks of the Vale of Pickering.	108
62.	Reconstructed cross section of Betton Farm South Quarry.	109
63.	Betton Farm South Quarry and its fauna.	110
64.	Map of the Pleistocene and underlying Jurassic strata at Filey Brigg (Itinerary 10).	112
65.	Lithic log of the Corallian at Filey Brigg.	113
66.	The Corallian sequence at Filey Brigg.	114
67.	*Thalassinoides* burrow systems in the Hambleton Oolite Member, Filey Brigg.	115
68.	View east along the cliffs on the south side of Filey Brigg.	116
69.	Glacial diamicton facies in the lower till at the cliff base at Wool Dale.	117
70.	Map of the Speeton section (Itinerary 11).	119
71.	Simplified lithic log of the D to lower B beds of the Speeton Clay Formation.	120
72.	Bed D1, the 'compound nodule bed', Speeton.	121
73.	Three distinctive marker horizons in the mid C beds, Speeton.	122
74.	Red Hole, Speeton.	125
75.	Locality map of the Flamborough area (Itineraries 12-14).	127
76.	Lithic log of the Chalk at Thornwick Bay and North Landing.	128
77.	The Chalk at Little Thornwick Bay.	129
78.	Thornwick Nab.	129
79.	View across Great Thornwick Bay.	130
80.	Marker flint bands at the east side of Great Thornwick Bay.	131
81.	Structures in the Chalk, West Cliff, Selwicks Bay.	133
82.	West Cliff, Selwicks Bay, looking due west.	134
83.	West Cliff, Selwicks Bay.	135
84.	Quaternary deposits on Flamborough Formation chalks, High Stacks.	136
85.	Lithic log of the Chalk from High Stacks to Sewerby Steps.	138
86.	Composite section of the glacial deposits at West Nook, South Landing.	141

		Page
87.	Lowest part of the glacial succession at West Nook recorded in Fig. 86.	142
88.	The three Danes Dyke Lower Marls, 50 m east of Danes Dyke.	143
89.	Quaternary deposits in west cliff, Danes Dyke.	144
90.	Diagrammatic section of the buried cliff at Sewerby.	147
91.	Glaciology of the eastern end of the Vale of Pickering.	150
92.	The northern end of the Barmston section.	152
93.	Mere deposits at Skipsea Withow Gap.	153
94.	Matrix colour variation within the Skipsea Till.	154
95.	Withernsea Till exposed in Withernsea north cliff.	157
96.	Tills at Dimlington High Land.	158
97.	Bedded sands above Skipsea Till, Dimlington High Land.	159

LIST OF TABLES

1.	Subdivision of the Lower Jurassic sequence.	8
2.	Subdivision of the Middle Jurassic sequence.	13
3.	Subdivision of the Upper Jurassic sequence.	14
4.	Subdivision of the Lower Cretaceous sequence.	16
5.	Subdivision of the Chalk Group sequence (Upper Cretaceous).	17
6.	Subdivision of the Quaternary sequence.	18

1 GEOLOGY OF THE YORKSHIRE COAST

P.F. Rawson and J.K. Wright

1.1 Introduction

The Yorkshire coast provides magnificent exposures of Jurassic and Cretaceous rocks that were deposited in the Cleveland Basin and on the adjacent northern part of the East Midlands Shelf (Figs 1, 2). In addition much of the lower-lying ground is covered by a thick sequence of Quaternary deposits, well-exposed along the Holderness coast. Thus the coastal area is firmly established as a standard for comparison with both the less well exposed areas inland and also for the offshore North Sea basins. It has attracted the attention of geologists from the earliest days of our science and the appearance of this fourth edition of our guide effectively celebrates 200 years of published research on the area. William Smith visited several times in the first two decades of the nineteenth century and recognised almost all the groups of strata that he had previously defined in southern England. Largely through his nephew John Phillips, and particularly after he had settled near Scarborough in 1828, Smith encouraged the publication of descriptions of the fascinating series of rocks he had found, although local workers had already become involved. The Reverend George Young of Whitby had been a student at Edinburgh where Playfair had aroused his interest in geology. Hence when Young published his *History of Whitby* in 1817, he included an outline account of the local geology. Then in 1822, he and local artist John Bird published *A Geological Survey of the Yorkshire Coast*. A revised edition was issued in 1828, followed shortly afterwards by John Phillips' *Illustrations of the Geology of Yorkshire; or, a description of the Strata and Organic Remains of the Yorkshire Coast* (1829). These pioneering works laid a firm though sometimes conflicting foundation for later researchers to build upon.

Our understanding of the geology of north and east Yorkshire was substantially improved during the detailed mapping of the area by officers of the Geological Survey in the eighteen seventies and eighties. The Jurassic part of this work was elegantly summarised by Fox-Strangways (1892), a volume which has provided a firm foundation for all future work.

Research has continued ever since and, as geological concepts and techniques have evolved, successive generations of geoscientists have returned to the classic exposures here, so that significant advances have been made since the appearance of the last edition of this guide in 2000. In addition a huge seismic database is now freely available online so that several seismic sections are included to illustrate aspects of the tectonic history of our area.

Field procedure and safety
The Yorkshire coast is renowned for its rugged beauty. It can also be very dangerous and it is constantly changing. Major landslips continue to occur, in some cases rendering access to well-known sections difficult or impossible, and the Holderness till cliffs are continuously retreating, at a rate of up to 2 metres per year. Users of this guide are advised to check for possible access problems and, before following the coastal itineraries, consult the local tide tables as there is a constant risk of being cut off by the tide. Unless an access or escape route is immediately adjacent, **never start work on a rising tide**. The cleanest exposures are often at the cliff foot where there are frequent landslips and an ever-present danger of falling rock; take sensible precautions by **keeping away from the cliff foot** as

1 Geology of the Yorkshire Coast

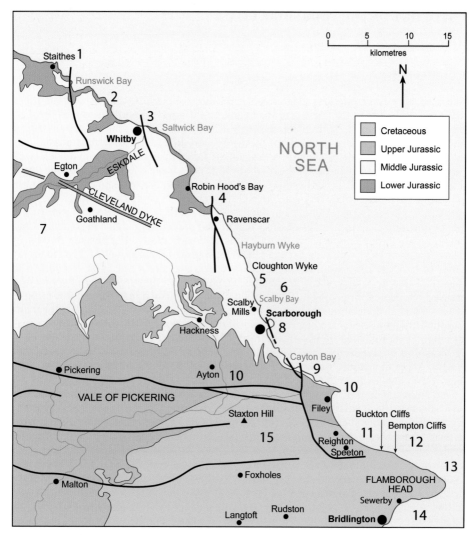

Figure 1. Outline geological map with itineraries: northern area.

far as possible, avoiding areas of recent cliff fall. Safety helmets are essential for work anywhere near the cliff foot, while safety glasses are also necessary when hammering. The shore is often rocky or boulder strewn and some of the rocky areas are very slippery —especially where shale scars are covered in an algal slime. Accidents happen, so ensure that someone always knows where you will be visiting and when.

It is also important that visitors help to protect and conserve the exposures described here. A huge number of parties and individuals visit the coast and adjacent moorlands, creating considerable pressure on conservation bodies and landowners. Some of the inland exposures described here are in the North York Moors National Park, while many of the coastal sections lie within either the North Yorkshire and Cleveland Heritage Coast

1 Geology of the Yorkshire Coast

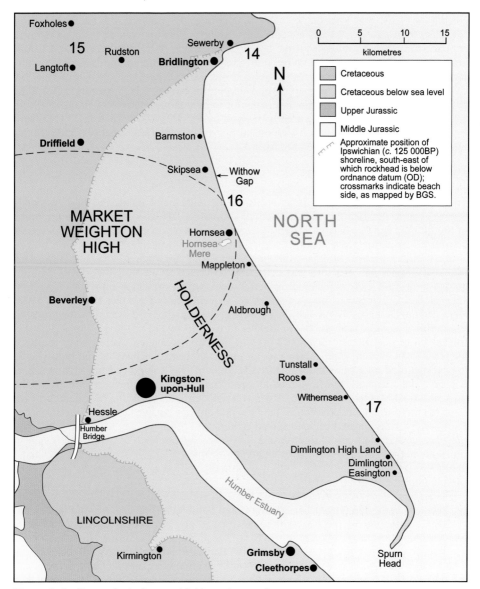

Figure 2. Outline geological map with itineraries: southern area.

(Saltburn to Scalby Mills) or the Flamborough Head Heritage Coast (Reighton to Sewerby). Some of the coastline is owned by the National Trust, who have created pathways along and down the cliffs: please do not stray from the marked routes and try to avoid disturbing nesting seabirds.

Please also bear in mind that there has been serious over-collecting (especially of fossils) from many of the best-known coastal sites. We recommend that users of this guide

1 Geology of the Yorkshire Coast

consult the Geologists' Association Code for Geological Fieldwork. It is rarely necessary to collect fossils *in situ*; excellent specimens can be picked up loose from fallen nodules or patches of shingle at many of the localities described. Note also that several localities are Sites of Special Scientific Interest (SSSIs) where hammering of the solid rock is forbidden.

Appropriate Ordnance Survey (OS) and Geological Survey (GS) maps are listed at the beginning of each itinerary. Location maps for individual itineraries are simplified from 1:25 000 OS maps. The whole area is covered by the 1:250 000 OS Road 4 (Northern England) map, while the British Geological Survey's Tyne-Tees Sheet (1:250 000) embraces the area from Bridlington northwards.

1.2 The structural framework

Towards the end of Triassic (Rhaetian) times the Cleveland Basin and East Midlands Shelf began to develop through differential subsidence, which continued through much of the Jurassic and Cretaceous. Their Mesozoic configuration may reflect the buried Carboniferous structure (Kent 1980b).

The northern part of the East Midlands Shelf forms the Market Weighton High. This regionally important structure was originally described as an anticline, but is now regarded as a rigid E-W orientated unfolded block which remained buoyant throughout Jurassic and Cretaceous times (supported by a deeply buried granite: Bott *et al.*, 1978; Kent, 1980a; Donato, 1993) while the Cleveland Basin to the north was subsiding rapidly. For much of Jurassic time the high probably formed an area of shallow water deposition, though those post-Liassic sediments which originally extended over it were removed by pre-Albian erosion (see Tables 1–6 for stage/age terms). However, the high should not be regarded as a permanent feature. At times it was submerged and, for instance, deeper-water sediments of Mid Callovian age occur at outcrop on the northern edge of the high in the Kirby Underdale area, 20 km west of Driffield. From the Middle Oxfordian onwards there was no expression of the high and argillaceous sediments (Ampthill Clay, Kimmeridge Clay) appear to have spread right across from the East Midlands Shelf into the Cleveland Basin without interruption. Conversely, from Mid Volgian to Early Albian times the high appears to have been emergent, forming a barrier between the Cleveland Basin and the shelf to the south, until it was finally submerged again by the Mid Albian marine transgression.

The high is essentially a hinge between the shelf and the Cleveland Basin, the main line of inflection being the Howardian-Flamborough Fault Belt (Fig. 3). Faulting in this belt probably started during the Carboniferous, when it may have been contiguous with the Craven Fault Belt some 140 km to the west. The Jurassic rocks of the Howardian Hills are intensively disturbed by a series of E-W faults, which pass eastwards beneath the Chalk (Kirby & Swallow, 1987). They probably originated as Jurassic growth faults and the first phase of movement was in mid-Oxfordian times (Wright, 2009). They were then intermittently active during the latest Jurassic and Early Cretaceous interval (Late Cimmerian movements). Thus, at times, a submarine or subaerial fault scarp probably formed the northern boundary of the Market Weighton High. Further reactivation occurred during and after deposition of the Chalk, indicated by E-W trending faults and 'shatter belts' in the Chalk. This movement was another local response to more widespread events, reflecting both the Laramide movements (end-Cretaceous to Paleocene) and the Alpine orogeny (Oligocene–Miocene), and embraced both tensional and compressional phases (Starmer, 1995a; Itineraries 13, 15).

1 Geology of the Yorkshire Coast

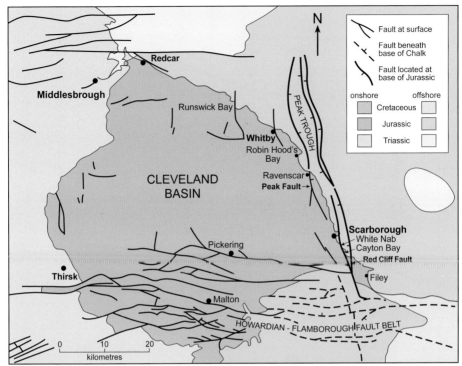

Figure 3. The structural framework. After Kirby and Swallow (1987); Milsom and Rawson (1989); BGS (1998b) and Wood (this volume).

While the Howardian-Flamborough Fault Belt formed the southern margin of the Cleveland Basin (Fig. 3), its western limit was defined by the Pennine High and its northern by the Mid North Sea High. A number of faults are known within the basin, especially along the coast where several N-S trending ones occur (Fig. 3; Powell, 2010, fig. 3). The best known is the Peak Fault (Itinerary 4), whose origin has long been debated. Seismic information shows that it forms the western boundary to a narrow N-S trending graben, the Peak Trough, which runs obliquely to the coast with the Red Cliff Fault forming its eastern margin (Milsom & Rawson, 1989; British Geological Survey, 1998a, 1998b). Fault movement in the trough probably occurred intermittently from Triassic to Miocene times and influenced sedimentation particularly during the latest Early Jurassic to Mid Jurassic interval.

The whole region was gradually uplifted from very late in the Cretaceous through much of the Cenozoic (Bray *et al.*, 1992). The uplift was accompanied by inversion of the Cleveland Basin to form the present day E-W trending Cleveland Anticline (Fig. 4). The main phase of inversion is assigned either to the Late Cretaceous and Early Cenozoic (Kent, 1980a) or to the Cenozoic alone (Hemingway & Riddler, 1982). This inversion mainly reflects compression from the south (Alpine movements). The axes of several subsidiary domes and troughs are aligned obliquely to the main axis (Fig. 4), but the most spectacularly exposed of these minor fold structures, the Robin Hood's Bay Dome on the coast south-east of Whitby (Itinerary 4; Figs 11, 28), may have formed in response to the swelling of Zechstein salt beneath (Section 2 below).

1 Geology of the Yorkshire Coast

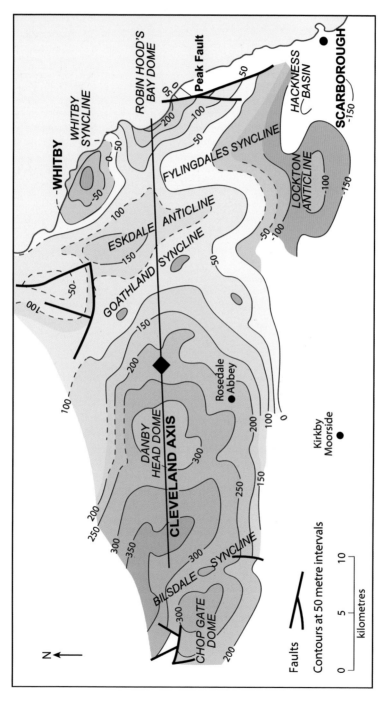

Figure 4. Structural inversion of the Cleveland Basin. Based on contours drawn on the top surface of the Dogger Formation. Modified from Kent (1980b, fig. 24) and Hemingway and Riddler (1982).

1 Geology of the Yorkshire Coast

1.3 Stratigraphy, Palaeogeography and Environments

Sea-level rise in latest Triassic to earliest Jurassic times established a marine regime over the region, which was marginal to the Southern North Sea Basin (Fig. 5A). By the beginning of the Jurassic the area lay at about 30° north of the equator and enjoyed a sub-tropical climate. Shorelines were generally some distance from present outcrops and the sequence consists predominantly of offshore argillaceous sediments, though shallow-water, more sandy sediments also occur. The Lower Jurassic sediments are placed in the Lias Group, the nomenclature of which has been revised by Knox (1984), Powell (1984) and Howard (1985); five formations are recognised (Table 1), while many of the long-established smaller, essentially lithological, subdivisions have been retained and formalised as members. The whole succession was deposited in two upward-coarsening cycles; the lower embracing the Redcar Mudstone to Staithes Sandstone formations and the upper the Cleveland Ironstone to Blea Wyke Sandstone formations (Powell, 2010). Smaller-scale cyclicity is also apparent at many levels, sometimes reflecting alternation of storm-dominated and quiet-water conditions.

The Calcareous Shale Member (Redcar Mudstone Formation) is the lowest unit to be seen in detail in the itineraries to be described in this guide. It consists of silty, calcareous mudstones with frequent calcareous concretions and shell beds. The overlying Siliceous Shale Member embraces silty shales alternating with thin, harder, fine-grained sandstones, the latter forming the harder ledges picked out by the sea in the intertidal zone at Robin Hood's Bay (Itinerary 4). Sellwood (1970) originally suggested that each sandstone band formed the top of a small-scale (1–4 m) coarsening-upward cycle, which he attributed to variation in sea level. However, the careful logging of the sequence by Hesselbo and Jenkyns (1996) showed that there are upwards of 70 of these sandy incursions varying considerably in their thickness and in the degree to which scouring was involved in their deposition. The suggestion of Van Buchem and McCave (1989) that the whole sequence was laid down in a fairly shallow marine shelf subject to periodic storm conditions and that the sandy beds represent proximal to distal storm beds deposited by storms of widely varying intensity, is much more convincing (Powell, 2010).

This long interval of shallow marine, storm-influenced sedimentation was terminated abruptly by the basal Pliensbachian sea-level rise which restored deeper water hemipelagic mud sedimentation to the basin (Pyritous Shale and Ironstone Shale members). At times the sea floor became anoxic, but on other occasions *Pinna*, a large, semi-infaunal bivalve, thrived here. Later Pliensbachian times saw a shallowing of the sea and argillaceous sands and silts were deposited to form the Staithes Sandstone Formation (Itinerary 1). Storm-influenced sedimentation again played an important role at this time (Howard, 1985; Powell, 2010). Much of the sediment was derived from a Pennine landmass to the north-west. A shoreline in that direction is also indicated by the overlying Cleveland Ironstone Formation, deposition of which commenced in response to a minor sea-level rise. The formation consists of a nearshore sequence in Cleveland dominated by oolitic ironstones, passing to the south-east into a more argillaceous, deeper-water succession composed of a series of minor coarsening-upward cycles. Each cycle is capped by an ironstone; the iron was presumably leached from a low-lying, subtropical, well-vegetated landmass.

The Toarcian saw important changes in the pattern of sedimentation in the Cleveland Basin, marked initially by a new phase of mudrock sedimentation, the Whitby Mudstone Formation, which has a sharp, though conformable, basal contact with the Cleveland Ironstone Formation. The initial Grey Shale Member consists of silty, micaceous mud-

1 Geology of the Yorkshire Coast

STAGE/SUBSTAGE	CHRONOZONE	LITHOSTRATIGRAPHY		
174 Ma	*Pleydellia* Aalensis			
TOARCIAN	*Dumortieria* Pseudoradiosa	Blea Wyke Sandstone Formation	Yellow Sandstone Member	9 m
	Phylseogrammoceras Dispansum		Grey Sandstone Member	9 m
	Grammoceras Thouarsense	Whitby Mudstone Formation	Fox Cliff Siltstone Member	11 m
	Haugia Variabilis		Peak Mudstone Member	13 m
	Hildoceras Bifrons		Alum Shale Member	37 m
	Harpoceras Serpentinum		Mulgrave Shale Member	32 m
183 Ma	*Dactylioceras* Tenuicostatum		Grey Shale Member	14 m
UPPER PLIENSBACHIAN (DOMERIAN)	*Pleuroceras* Spinatum	Cleveland Ironstone Formation	Kettleness Member	10 m
	Amaltheus Margaritatus		Penny Nab Member	19 m
LOWER PLIENSBACHIAN (CARIXIAN)	*Prodactylioceras* Davoei	Staithes Sandstone Formation		25 m
	Tragophylloceras Ibex		Ironstone Shale Member	63 m
191 Ma	*Uptonia* Jamesoni		Pyritous Shale Member	26 m
UPPER SINEMURIAN	*Echioceras* Raricostatum	Redcar Mudstone Formation	Siliceous Shale Member 39 m	
	Oxynoticeras Oxynotum			
	Asteroceras Obtusum			
LOWER SINEMURIAN	*Caenesites* Turneri			
	Arnioceras Semicostatum		Calcareous Shale Member 127 m	
199 Ma	*Arietites* Bucklandi			
HETTANGIAN	*Schlotheimia* Angulata			
	Alsatites Liassicus			
201 Ma	*Psiloceras* Planorbis			

Table 1. Subdivision of the Lower Jurassic (Hettangian to Toarcian) sequence.

1 Geology of the Yorkshire Coast

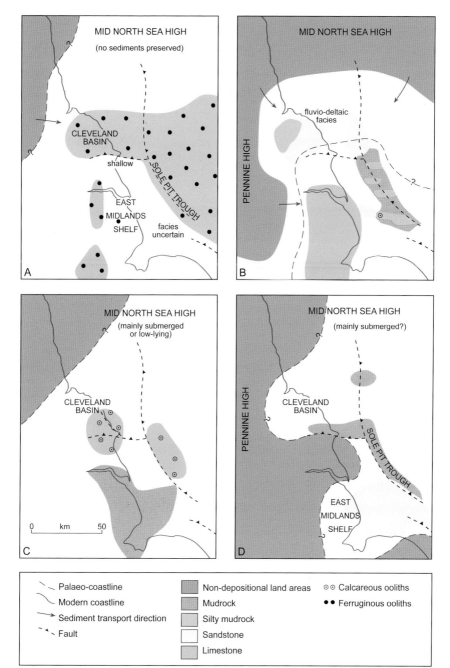

Figure 5. The palaeogeographical setting. (A) Early Jurassic, (B) Mid Jurassic, (C) Late Jurassic, (D) Early Cretaceous. Simplified from maps J2b, J5, J9 and K1 in Cope *et al.* (1992).

1 Geology of the Yorkshire Coast

stone. Through most of this unit, a rather sparse, normal marine fauna occurs, reflecting sedimentation in a relatively well-oxygenated environment. Marked faunal and sedimentological changes then become apparent, reflecting a global Early Toarcian Oceanic Anoxic Event (OAE). Over a period of 300 kyr there was one of the largest negative carbon isotope excursions of the Phanerozoic (Cohen *et al*., 2007). This is thought to have been caused by the injection of huge amounts of carbon dioxide into the atmosphere and oceans during major volcanic episodes in Africa (Palfy & Smith, 2000; McElwain *et al*., 2005; Svenson *et al*., 2007). There was a marked increase in global temperature (Rosales *et al*., 2004) and world-wide hot, humid conditions, with extensive run-off of nutrients from continents, saw huge amounts of organic matter deposited in the ocean basins, which became anoxic. As a result, a major extinction event occurred. The highest part of the Grey Shale Member marks the extinction interval (Danise *et al*., 2015).

The overlying Mulgrave Shale Member is a finely laminated mudrock that shows clear evidence of oxygen depletion in the bottom waters. There is a very restricted and specialised bottom-living fauna (e.g. Caswell *et al*., 2009), especially in the lower part (Exaratum Subzone) which, together with the extinction interval at the top of the Grey Shale Member, marks the "interval of maximum restriction" (Danise *et al*., 2015). Here the epifaunal, suspension-feeding opportunistic bivalve *Pseudomytiloides dubius* is common and was apparently adapted to tolerate low-oxygen conditions, sometimes attached to strands of seaweed for at least part of its life cycle. In contrast, there is quite a diverse fauna of free-swimming animals (ammonites, fish, reptiles, etc.). The main water mass must therefore have been well oxygenated while the sea floor was generally oxygen-low or deficient. Such shales were deposited over much of Europe (Posidonienschiefer, Schistes Carton, etc.). They form a source rock for oil in North Germany, the Netherlands and the Paris Basin, and when freshly broken the Yorkshire sediments smell strongly of oil. The lower part of the Mulgrave Shale Member ('Jet Rock' *sensu stricto*) is more finely and persistently laminated than the higher part ('Bituminous Shales') and represents peak oxygen reduction. Jet itself occurs as very thin, elongate coal-like seams, which are flattened, drifted logs of wood allied to the modern araucarians.

The Alum Shale Member marks a return to more 'normal' marine mudstone sedimentation with some bottom-living organisms, including abundant examples of the mud-burrowing bivalve *Dacryomya ovum*. Rare trace fossils occur, including *Thalassinoides*.

Higher Toarcian rocks are preserved only in small pre-Dogger (Mid Jurassic) synclines and are best exposed along the coast immediately to the south-east of the Peak Fault. Here, the shallower-water sediments above the Alum Shale Member form a sequence of three fining-upwards cycles (the lowest forming the Peak Mudstone Member and the other two the Fox Cliff Siltstone Member) overlain by two coarsening-upwards cycles (Grey Sandstone and Yellow Sandstone members). Individual cycles vary between 9 and 12.6 m maximum thickness (Knox, 1984).

Prior to deposition of the Dogger Formation, the Cleveland Basin was uplifted, gently folded and subject to pericontemporaneous faulting, resulting in the removal of up to 60 m of Liassic sediment, as seen at Whitby and Ravenscar (Itineraries 3 and 4). The marine Dogger Formation thus rests with a marked unconformity on the Liassic strata in this area. The presence of *Arenicolites* 'U' tubes burrowed into the Alum Shale Member shows that the latter was still relatively unconsolidated despite the 4 myr break here. The Dogger is a complex unit, ferruginous throughout. In the west, varied sequences of shallow-marine strata have been traced (Hemingway, 1974a; Powell, 2010). However, at Whitby and Ravenscar only 0.5–1 m of ferruginous sandstone full of eroded fragments of Whitby Mudstone nodules is seen.

1 Geology of the Yorkshire Coast

The latest Toarcian events represent the early stages of a regional uplift which created an extensive landmass over much of the central and northern North Sea. This culminated at the beginning of the Aalenian (Middle Jurassic). There followed a world-wide reduction in sea level and a radical change in regional palaeogeography (Fig. 5B). The shallow shelf situated marginal to the Mid North Sea High was initially marine (Dogger Formation) but soon became subject to periodic incursions of deltaic and fluvial sediments derived from the newly uplifted land area so that sequences of marine and non-marine strata regularly alternate (Ravenscar Group). In contrast, much of central and southern England was covered by a warm, shallow shelf sea during this time.

The depositional environment of the non-marine beds of the Ravenscar Group has caused much discussion. The initial estuarine hypothesis of Fox-Strangways (1892) has been superseded and the debate now centres on deltaic versus alluvial plain hypotheses. In fact, the Ravenscar Group shows features characteristic of both. The overall evidence is that following each of the marine intervals—Dogger Formation, Eller Beck Formation, Lebberston Member and Scarborough Formation, there was a rapid development of small, prograding deltas, ultimately coalescing into a large alluvial plain. There is no evidence for a large river system feeding into a large delta.

The Dogger Formation is itself penetrated by plant roots from the overlying non-marine Saltwick Formation. In the latter, grey mudstones and siltstones predominate, with intercalations of laterally impersistent channel sandstones, reflecting the establishment of fluvial, deltaic and brackish lagoon conditions. Plants and trees flourished, and numerous fossil footprints show that the area was colonised by dinosaurs of both plant-eating and carnivorous varieties (Romano *et al*, 1999; Whyte *et al*., 2010).

During the marine invasion of the Eller Beck Formation, the sea transgressed from the north-east; the formation on the coast consisting of marginal-marine sandy and ferruginous sediments with a 'dwarfed' bivalve fauna and starfish resting traces.

The Cloughton Formation is a complex unit, marking a period when initial non-marine conditions (Sycarham Member) became established, to be replaced by a partial return to marine conditions (Lebberston Member) succeeded by non-marine conditions again (Gristhorpe Member). Thus, the Sycarham Member consists of overbank coastal plain muds with minor sand channels. These are overlain by stacked, erosive-based channels full of fluvial and deltaic sands.

The Lebberston Member marks a return to fully marine conditions in the south only, ooidal limestones (Yons Nab, Itinerary 9) passing northwards into shelly carbonates (Millepore Bed – Osgodby Point, Itinerary 9) and sandy, shelly ferruginous beds at Cloughton (Itinerary 5). These beds are then overlain by the more marginally marine Yons Nab Beds. The Gristhorpe Member marks a return to lower delta-plain environments, with abundant plant remains as a diverse flora colonised the substrate.

The Scarborough Formation marks a major period during the accumulation of the Ravenscar Group when fully marine conditions spread from the east right across the Cleveland Basin. Consisting of a complex interplay of marine sands, clays and ferruginous carbonates, deposited in a sandy, littoral embayment passing offshore into calcareous mud, the frequent occurrence of ammonites shows the fully marine nature of the succession and enables the formation to be firmly dated as mid Bajocian.

The Scalby Formation rests with marked erosion on the Scarborough Formation, the sheet sands of the basal Moor Grit Member filling channels eroded in the uppermost members of the Scarborough Formation. The Moor Grit and the basal part of the overlying Long Nab Member have yielded very sparse dinoflagellate floras of latest Bajocian to Bathonian age. Dinoflagellates from the upper Long Nab Member at Newtondale indicate

1 Geology of the Yorkshire Coast

a Bathonian age (Riding & Wright, 1989). The evidence for a major break in the succession either beneath or above the Scalby Formation (Nami & Leeder, 1978) appears slim. The Long Nab Member, with its meandering streams, superbly seen in the meander belt at Scalby Bay (Itinerary 6), flood plains with desiccation cracks and dinosaur footprint trails, fossil soils with rootlet beds and the extensive development of sphaerosiderite all point to an alluvial origin.

The Callovian Stage marks the return of fully marine conditions to the Cleveland Basin. Callovian and also Oxfordian sediments in the basin often show significant facies differences to those on the East Midlands Shelf. The effects of uplift of the intervening Market Weighton High can be difficult to determine however. Absence of strata at Market Weighton may be because they were never deposited there, or it may be that they were eroded away by the subsequent uplift. Facies changes as one approaches the high are key.

The Lower Callovian shelly limestones of the Fleet Member of the Abbotsbury Cornbrash Formation and the Cayton Clay Formation are very similar in lithology across England, including Yorkshire, and were not affected by any uplift (Page, 1989). In contrast, the succeeding fully marine sands of the Red Cliff Rock Member of the Osgodby Formation pass into marginally marine facies on the northern edge of the High (Powell *et al.*, 2018) and are not present south of Market Weighton (Page, 1989). The Red Cliff Rock thus cannot be correlated with the Kellaways Sand Member of Wiltshire as previously thought (Hemingway, 1974a). Much of the Kellaways Formation and of the lower part of the Peterborough Member, the lowest member of the Oxford Clay Formation, are absent in the Cleveland Basin, though a section of the middle part of the Peterborough Member did extend as far north as Malton (Wright, 1968b).

The grouping together into the Osgodby Formation of the three shallow marine sandy Callovian members (Red Cliff Rock, Langdale and Hackness Rock members) masks an important stratigraphic break in the Cleveland Basin (Table 2), a gap of some half million years beneath the Langdale Member not being represented by strata (Wright, 1968a). Eventually, a connection with sedimentation on the East Midlands Shelf developed and the two lowest members of the Oxford Clay Formation, well developed there, are represented in part in the Cleveland Basin by sandy, shallow-marine beds. The upper part of the Peterborough Member is the equivalent of the sands and silts of the Langdale Member, the ubiquitous occurrence of the ammonite *Erymnoceras* showing the connection, and the clays of the Stewartby Member are represented in the north-east by the Hackness Rock Member, a condensed, iron-rich deposit which frequently yields excellently preserved ammonites (Callomon & Wright, 1989).

By the beginning of the Oxfordian, the finer-grained sediments of the Weymouth Member of the Oxford Clay Formation spread over the whole area at least as far north as Scarborough. Little true clay is present, much of the Weymouth Member here consisting of argillaceous siltstone. Oxford Clay sedimentation did not last for long before the basin began to fill up and the calcareous sands (grits) and limestones of the Corallian Group accumulated (Fig. 5C). These are equivalent to the highest Oxford Clay and much of the Ampthill Clay formations of the East Midlands Shelf, and indicate a second phase of inversion of basin and shelf. The Corallian Group is divided into three formations (Lower Calcareous Grit, Coralline Oolite and Upper Calcareous Grit formations, Table 3); the name of the middle formation, the Coralline Oolite, stemming from William Smith (Smith, 1829–30). The rapid lateral and vertical changes within the Corallian Group have led to the recognition of a dozen or more local members (summarized in Wright, 1972, 1983, 2009: see also Fig. 61). The Corallian facies in general indicate a warm, shallow, well-oxygenated sea in which ooidal carbonate banks and coral reefs developed. Fine-

1 Geology of the Yorkshire Coast

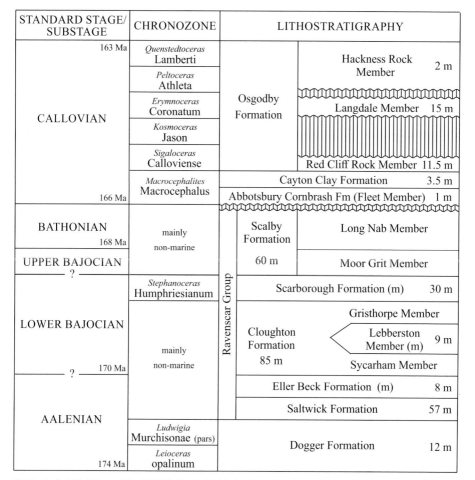

Table 2. Subdivision of the Middle Jurassic (Aalenian to Callovian) sequence.

grained (micritic) limestones accumulated in areas offshore to the shelf and in sheltered, back-reef lagoons. Occasional uplift of source areas saw several incursions of quartz sand ('calcareous grits') into the area.

The Lower Calcareous Grit Formation is predominantly a *Rhaxella* spiculite containing a small proportion of fine quartz sand (Wright, 2009, fig. 4). The sand was probably derived from the northern margin of the Market Weighton High (Wright, 1983). The siliceous spicules were readily dissolved and are the source of diagenetic cement which is often concentrated into chert nodules or bands. The settled, quiet conditions necessary for the prolific growth of sponges seem to have suited ammonites very well and the calcareous concretions of this formation have yielded some of the best-preserved examples found in Britain (Wright, 1983; Wright et al., 2014).

The transition from sand deposition to the carbonates of the Coralline Oolite Formation is shown in stages in the basal Yedmandale Member, with its upwards transition from beach sands through ferruginous carbonates and coral reefs into ooid shoals (Wright,

1 Geology of the Yorkshire Coast

STANDARD STAGE/ SUBSTAGE			CHRONOZONE	LITHOSTRATIGRAPHY	
145 Ma					
TITHONIAN	REGIONAL STAGE	VOLGIAN (pars)	*Pectinatites* Pectinatus (pars)	Kimmeridge Clay Formation	c 305 m
			Pectinatites Hudlestoni		
			Pectinatites Wheatleyensis		
			Pectinatites Scitulus		
			Pectinatites Elegans		
152 Ma			*Aulacostephanus* Autissiodorensis		
KIMMERIDGIAN			*Aulacostephanus* Mutabilis		
			Aulacostephanus Eudoxus		
			Rasenia Cymodoce		
157 Ma			*Pictonia* Baylei		
UPPER OXFORDIAN			*Amoeboceras* Rosenkrantzi	Ampthill Clay Formation	46 m
			Amoeboceras Regulare		
			Amoeboceras Serratum		
			Amoeboceras Glosense	Upper Calcareous Grit Fm (Corallian Group)	11 m
MIDDLE OXFORDIAN			*Cardioceras* Tenuiserratum	Coralline Oolite Formation (Corallian Group)	60 m
			Cardioceras Densiplicatum		
LOWER OXFORDIAN			*Cardioceras* Cordatum	Lower Calcareous Grit Fm (Corallian Group)	50 m
163 Ma			*Quenstedtoceras* Mariae	Oxford Clay Formation	37 m

Table 3. Subdivision of the Upper Jurassic (Oxfordian to Volgian) sequence.

1992). The increasing wave energy, warm, shallow seas and minimal clastic input then resulted in the accumulation of substantial amounts of oolitic and shelly limestones. During deposition of the Hambleton Oolite Member, thick sequences of cross-bedded, shallow water oolites occupied the area right across from Scarborough to the Hambleton Hills. Southwards, further uplift of the Market Weighton High saw an influx of fine quartz sand (Birdsall Calcareous Grit Member) dividing the Hambleton Oolite into lower and upper leaves.

Uplift of source areas to the north then led to the sweeping in of the fine-grained siliciclastic sediment of the Middle Calcareous Grit Member, which eventually spread right down into the basin as far as Malton (Wright, 2009). With the waning of this influx, ooidal sedimentation (Malton Oolite Member) spread across the whole basin north of Market Weighton. Ooid shoals were only localized, much of the basin consisting of an offshore shelf accumulating ooidal micrites in which are found giant perisphinctid ammonites (the 'snake beds' of the quarrymen). Near Scarborough, a substantial coral reef developed (Wright & Rawson, 2014, Itinerary 10), eventually to be overwhelmed as ooidal sands prograded across much of the Cleveland Basin.

A period of localized uplift and erosion was then followed by uniform submergence of the whole basin and the deposition of the Coral Rag Member, a shallow, basin-wide accumulation of micritic limestone with a prolific coral fauna (Wright, 2009), centred on a lagoon accumulating carbonate muds near Pickering.

Progressive uplift of the source areas to the north then brought an end to the reefs as red, lateritic clay was washed in. The clay is then succeeded by the siltstones, fine-grained sandstones and silicified *Rhaxella* spiculites of the Upper Calcareous Grit Formation (Wright, 1996).

In late Oxfordian times, Ampthill Clay facies spread northwards beyond Market Weighton into the Vale of Pickering (Cox & Richardson, 1982). None of the exposures and borings on the north side of the vale show any sign of shallowing facies and the whole Cleveland Basin seems to have been submerged beneath a muddy sea. This continued into the Kimmeridgian and early Volgian, with the accumulation of a thick sequence of Kimmeridge Clay. Formerly known only from scattered brick pits and minimal coastal exposures, the full sequence has been penetrated in a succession of boreholes drilled in the Vale of Pickering (Herbin *et al.*, 1991). The succession is complete until early Pectinatus Zone or early Pallasioides Zone times, but uppermost Jurassic and lowermost Cretaceous strata are absent (Tables 3 and 4), there being a time gap of some 9 myr, during which the sea probably remained over the area (Rawson & Riley, 1982). Cretaceous sedimentation eventually commenced again with deposition of the Lower Cretaceous Speeton Clay Formation (Table 4). Offshore, over much of the North Sea Basin, sedimentation continued during this interval to produce a thick sequence of organic-rich muds that form the source rock for most of the North Sea oil.

The Speeton Clay Formation accumulated to the north of an emergent Market Weighton High (Fig. 5D) which separated it from the more varied sediments (sandstones, ironstones, limestones and mudrocks) of the East Midlands Shelf (Rawson, 2006). It crops out as faulted inliers in the Vale of Pickering and in a strip along the Wolds escarpment to reach the coast south of Filey, where it is well exposed at the type section at Speeton (Itinerary 11). South of the outcrop, the Speeton Clay may reach up to a kilometre thick locally in the Flamborough Fault Zone (Kirby & Swallow, 1987) before disappearing southward across the Market Weighton High, but at Speeton the clays are only about 100 m thick. However, this section provides a unique sequence through almost the whole British marine Lower Cretaceous; the lower (and better preserved) part of the section correlating with the non-marine Wealden facies of southern England. The faunal and floral successions provide a standard for comparison with the much thicker North Sea and North German successions. Thin volcanogenic mudstones at several levels (Knox, 1991) reflect the activity of distant volcanoes, possibly in the Dutch sector of the southern North Sea.

The Speeton Clay Formation passes up into a distinctive thin red limestone—the Hunstanton Formation ('Red Chalk')—deposited as rising sea levels flooded the Market Weighton High again. As clastic input diminished a pure white limestone, the Chalk, began to accumulate, composed of calcareous algal plates (coccoliths) deposited as copepod faecal pellets in a clear warm sea that eventually extended over almost the whole of Britain and adjacent areas. A striking feature of some of the Yorkshire Chalk compared

1 Geology of the Yorkshire Coast

	STANDARD STAGE/SUBSTAGE				LITHOSTRATIGRAPHY			
	CENOMANIAN (pars)		101 Ma		Hunstanton Formation ("Red Chalk") 33 m			
LOWER CRETACEOUS	ALBIAN		113 Ma	M-U				
				L		A beds c. 12 m	UA1-UA4	c. 7 m
	APTIAN			U			LA1-LA6	5 m
			125 Ma	L	Speeton Clay Formation c.109 m	B beds c. 46 m	upper B beds	c. 9 m
	BARREMIAN			U			cement beds	c. 10 m
			129 Ma	L			LB1-LB6	27 m
	HAUTERIVIAN			U		C beds 39 m	C1-C7 mid	29 m
			133 Ma	L			C7 mid-C11	10 m
	VALANGINIAN			U		D beds 12 m	D1-D2D	1 m
			140 Ma	L			D2E-D8	11 m
	BERRIASIAN	REGIONAL STAGE		U				
				L				
			145 Ma					
U. J.	TITHONIAN		VOLGIAN		(Kimmeridge Clay Formation)			

Table 4. Subdivision of the Lower Cretaceous (Ryazanian to Albian) sequence.

with its southern England counterpart is its hardness; Jeans *et al*. (2014) showed that this reflected two phases of calcite sedimentation.

The Chalk Group forms the arcuate hills of the Wolds, reaching the sea at Flamborough Head to form a magnificent sweep of cliffs about 17 km long. The lower beds of the Chalk Group can be seen in Speeton (Itinerary 11) and Buckton cliffs, but the overlying sequence becomes increasingly difficult to reach and is inaccessible beyond Staple Nook, at the south-east end of Bempton Cliffs (Mortimore *et al*., 2001, p. 411). However, the famous contortion zone at Staple Nook (Starmer, 1995b) is readily seen from pleasure boats and together with the structure in Selwicks Bay (Itinerary 13) forms part of the Howardian-Flamborough Fault Belt. In the Flamborough region the beds become accessible again (Itineraries 12 to 14).

A major revision of the stratigraphy of the Chalk of the northern province (Yorkshire to North Norfolk was published by Wood and Smith (1978). The former division into Lower, Middle and Upper Chalk was unsatisfactory and the long established fossil zones were very vaguely defined. In modern lithostratigraphy, the Chalk Group at outcrop is divided into four formations (Table 5). The lowest, Ferriby Chalk Formation (20–30 m), is flintless. The Welton Chalk Formation (44–53 m) is characterized by nodular, burrow-form flints and the Burnham Chalk Formation (*c*. 130–150 m) by tabular and semi-tabular flint bands, especially in the lower part (Fig. 75) (Wood & Smith, 1978; Hildreth, 2018).

1 Geology of the Yorkshire Coast

The Flamborough Chalk Formation (260–280 m) is a flintless chalk with numerous very fine marl seams. Deep commercial wells in Holderness have shown that the Flamborough Chalk is overlain by flinty chalks up to 70 m thick (Sumbler, 1996), which offshore are assigned to the Rowe Formation.

Recognition of the lateral continuity of flint and marl bands has resulted in numerous named marker horizons. Thus there is a firm lithostratigraphic framework against which fossil occurrences can be calibrated (e.g. Whitham, 1991, 1993). Among the marker horizons, the 'Black Band' at the base of the Welton Chalk Formation (Itinerary 11) is the local representative of an Oceanic Anoxic Event (Schlanger & Jenkyns, 1976; Hart, 2018), while some of the overlying marl bands in the Welton and Burnham formations are believed to be of volcanogenic origin (Wray & Wood, 1998). Much of the Chalk sequence is demonstrated in Itineraries 11 to 15 and reviewed in detail by Mortimore et al., (2001).

No pre-Pleistocene Cenozoic sediments are known from the area, although sands and clays preserved in solution hollows in the Chalk of the Wolds were considered by Versey (1939) to be of this age. The burial history of the Cleveland Basin also indicates that some lower Cenozoic sediments may have been deposited, then removed during basin inversion later in the Cenozoic. Hemingway & Riddler (1982) deduced that as much as 1–1.25 km of Cenozoic sediments have been removed, though this is still a matter of contention.

Cenozoic igneous activity is represented by the Cleveland Dyke, a tholeiitic basalt intrusion peripheral to a volcanic province centred on the western Scottish island of Mull. The dyke extends for 250 km across southern Scotland and northern England and is visible on the North York Moors (Itinerary 7) but does not reach the coast. It has given K-Ar radiometric dates in the range of 56–59 Ma (Stone et al., 2012).

During the Pleistocene ice ages the area was glaciated at least twice and probably more often. In North Yorkshire only a few relics remain of deposits predating the last (Late Devensian) glaciation. The most extensive of these are thin patches of till ('boulder clay'

STAGE	BIOZONE	LITHOSTRATIGRAPHY		FORMER DIVISION
66 Ma MAASTRICHTIAN				
72 Ma CAMPANIAN 84 Ma	*Belemnitella mucronata* s.l.		Rowe Formation — ? —	Upper Chalk
	Gonioteuthis quadrata			
	Sphenoceramus lingua		Flamborough Chalk Formation > 300 m	
SANTONIAN 86 Ma	*Marsupites testudinarius*	White Chalk Subgroup		
	Uintacrinus socialis			
CONIACIAN 90 Ma	*Hagenowia rostrata*		Burnham Chalk Formation 150 m	
	Micraster cortestudinarium			
TURONIAN 94 Ma	*Sternotaxis plana*		Welton Chalk Formation 53 m	Middle Chalk
	Terebratulina lata			
	Mytiloides spp.			
CENOMANIAN (pars)	*Sciponoceras gracile*	Grey Chalk Subgroup	Ferriby Chalk Formation 33m	Lower Chalk
	Holaster trecensis			
	Holaster subglobosus			

Table 5. Subdivision of the Chalk Group sequence (Upper Cretaceous).

1 Geology of the Yorkshire Coast

or diamicton) scattered over the Vale of Pickering and the Tabular Hills and which probably derive from a small glacier that occupied the hills and then moved southwards into the vale (Powell et al., 2016). These may be of the same age as the more extensively developed, though generally poorly exposed in coastal sections, Basement Till of Holderness (Itineraries 14, 17). The Basement Till is known to underlie last interglacial beach deposits on the coast north of Bridlington at Sewerby (Table 6; Catt, 2007), though it may be older than the penultimate cold stage. Durable erratic rocks are known scattered across both the moors of North Yorkshire and the Yorkshire Wolds. It is unclear when they were deposited but they may relate to the most extensive glaciation of the UK, the Anglian (some 450,000 years ago).

Late Devensian deposits are more widespread, resulting from a late Devensian glaciation termed the Dimlington Stadial in the UK and named after the important type section in southern Holderness (Rose, 1985: Itinerary 17). At this time much of the higher land remained ice free, but ice passed down the Vale of York and the North Sea Lobe (NSL) of the ice sheet advanced south down what is now the western North Sea (Dove et al., 2017) to abut against the northern flanks of the North York Moors and adjacent coast (Fig. 6), where ice infilled the pre-glacial bays (e.g. the section at Upgang west of Whitby: Roberts et al., 2013). Meltwater flowed southward into the Vale of Pickering, and from ice margins both to the west and east, to form a large glacial lake, Lake Pickering (Fig. 6 and Itineraries 7 and 15: Evans et al., 2017) where lacustrine varved clays were deposited. Further south, ice spread over Holderness to abut against or slightly override a chalk

SYSTEM	SERIES/SUB-S.	MIS	BRITISH STAGE	SUB-STAGE	LITHOSTRATIGRAPHY
QUATERNARY (pars)	HOLOCENE	1	FLANDRIAN	0 Ka 11.5 Ka	Palaeo-mere sediments at Withow Gap and other Holderness sites (4.5-10.4 Ka)
QUATERNARY (pars)	UPPER PLEISTOCENE	2	DEVENSIAN	UPPER 31 Ka	silts, clays and sands (glacilacustrine/ glacideltaic deposits) at Barmston (11.3-15.0 Ka) Withernsea Till 'Dimlington Sands' (17.1 Ka) Skipsea Till Dimlington Silts (20.9 Ka)
QUATERNARY (pars)	UPPER PLEISTOCENE	3	DEVENSIAN	MIDDLE 58 Ka	silt matrix, chalk-clast gravels (gelifluction deposits) at Sewerby and Danes Dyke
QUATERNARY (pars)	UPPER PLEISTOCENE	4 5a-5d	DEVENSIAN	LOWER 116 Ka	Sewerby blown sand (89-102 Ka)
QUATERNARY (pars)	UPPER PLEISTOCENE	5e	IPSWICHIAN	128 Ka	Sewerby interglacial beach deposits
QUATERNARY (pars)	M. PLEIST. (pars)	?6 and older	PRE-IPSWICHIAN ('Wolstonian' and earlier)		Basement Till (incorporating rafts of Bridlington Crag)

Table 6. Subdivision of the Quaternary sequence.

1 Geology of the Yorkshire Coast

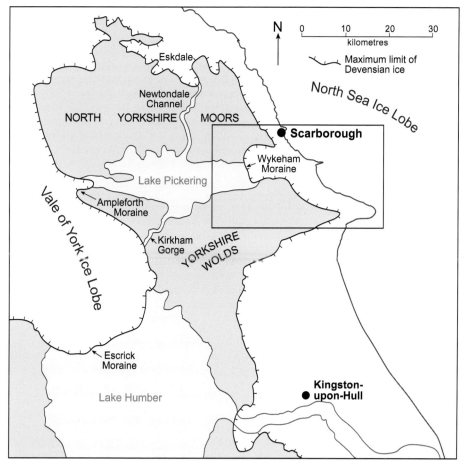

Figure 6. Palaeogeography during the Late Devensian (last) glaciation. Modified from Powell *et al.* (2016, fig. 1).

cliff line that extended from Sewerby to Hessle (Fig. 2) and across to Lincolnshire and possibly Norfolk. The main deposits left behind are tills and sand and gravel moraines (e.g. Catt, 2007; Evans & Thomson, 2010; Bateman *et al.*, 2015). The Skipsea Till covers the whole of Holderness, extends over the Chalk in the Flamborough area and infills the cores of bays further north. The Withernsea Till is of more limited extent in south-east Holderness (Bateman *et al.*, 2015, fig. 1) though it is also known at Filey Brigg (Boston *et al.*, 2010: Itinerary 10) and Robin Hood's Bay (Catt & Madgett, 1981). Both tills contain erratic pebbles and boulders derived from at least three distant sources—Scotland and the Lake District, north-eastern England and, much more rarely, Scandinavia (see Busfield *et al.*, 2015)—as well as local material. It is the subsequent erosion of this material that provides such a rich and fascinating suite of pebbles on our beaches—including the carnelians and agates so popular with collectors.

1 Geology of the Yorkshire Coast

1.4 The industrial story: past, present and future

There is considerable evidence of a long history of mining and quarrying in the area. In particular, the North York Moors and adjacent coastal area have been exploited extensively for a variety of resources (Goldring, 2001, 2006) and workings were often on a sufficiently large scale to permanently modify the landscape. The extractive industry continues to the present day and there are plans for future development too.

Jet from North Yorkshire has been used ornamentally since the Bronze Age, some 4000 years ago, but it is uncertain when mining first started. From the early 1850s a developing fashion for the use of jet in mourning jewellery led to the development of drift mines along both the coast (Itineraries 1-3) and inland in the North York Moors and Cleveland Hills, where spoil heaps are still visible. As fashions changed in the 1870s and cheap Spanish jet began to be imported the industry declined again.

Building stones, especially sandstones (particularly from the Ravenscar Group) and limestones (Corallian Group) have been quarried since Roman times. The results are visible in the remains of the Roman signal station at Scarborough, the 12th-century castle there, Whitby Abbey and other ecclesiastical buildings, and the many stone-built domestic buildings that give so much character to the Moors and adjacent areas. In the Wolds some of the hard chalks have been utilised, while in Holderness, erratic cobbles from the glacial tills were often laid in a distinctive herringbone pattern in the walls of cottages there. Many of these rocks, and basalt from the Cleveland Dyke, have also been used for roadstone, especially when parishes were responsible for local road maintenance. Corallian limestones are still quarried for aggregate and for agricultural use.

Middle Jurassic shales, Kimmeridge and Speeton Clays and some of the tills have been quarried for brick making, while for much of the 19th century high quality 'Roman' cement was produced in Hull from the calcareous concretions ('cementstones') occurring in the Alum Shale Member (Itinerary 3) and the Speeton Clay Formation (Itinerary 11). Glacial sands and gravels are still worked in the Vale of Pickering and Holderness.

From the earliest 1600s onward the alum industry developed on a very extensive scale. This worked the Alum Shale Member inland in the Cleveland Hills and particularly along the coast from Boulby to Ravenscar. The remains of many large quarries are still visible (e.g. Itineraries 3 and 4) along with tip heaps composed largely of the distinctive orange-coloured burnt shale. The last workings closed in 1871. Pybus and Rushton (1991) provide a useful summary of the history of the industry.

Thin seams and lenses of poor quality coal occur in the Ravenscar Group (Saltwick and Cloughton formations). These were mined in the North Yorkshire Moors from about 1640 till the end of the First World War and the remains of both bell pits and more extensive shaft and tunnel workings can still be traced. Thomas (2013–2015) has reviewed in detail the history and extent of mining, noting that the coal was used primarily for burning lime in kilns, though also for fuel.

Mining of the Cleveland Ironstone Formation started in 1837 at Grosmont (Owen, 1979) and workings extended from the Cleveland Hills in the north-west across the northern part of the moors and along the coast—where ironstone concretions had been collected from the shore since the late 18th century. By the 1870s this was the biggest iron mining area in Britain (Hemingway, 1974b) and iron was exported to many parts of the world. The last mine closed in January 1964. Evidence of ironstone extraction is seen in many areas, including Staithes (Itinerary 1). There are numerous booklets and short articles on individual mines available, many listed by Goldring (2001) who also gave a detailed account of the method of mining.

Dogger Formation ironstone was also mined on a small scale (Hemingway, 1974b), while a localised but very rich deposit of Middle Jurassic ironstone was mined in Rosedale from 1856 to 1926 (Hayes & Rutter, 1974).

In the 1930s, drilling for hydrocarbons commenced in the North York Moors and gas was first found in Eskdale in 1938. Subsequent successful finds have been made further south near Ebberston and in the northern part of the Vale of Pickering, where production is mainly from the Kirkham Abbey Formation (Permian). Gas from these areas has been utilised at Knapton Generating Station since 1995. There are now plans to frack for natural gas in one of the wells, at Kirby Misperton (Hughes *et al.*, 2016). Future developments in hydrocarbon exploration across both North and East Yorkshire may result from the recognition of the extensive distribution of the potentially hydrocarbon-rich Bowland Shale (Carboniferous): several companies have been awarded licences to explore much of north and east Yorkshire.

Another major development in the region is the mining of Permian salts from deep beneath the surface. In 1973 a mine opened at Boulby (Itinerary 1), to extract potash, and it now extends for some 5 km eastward beneath the North Sea. In 2016, Sirius Minerals were granted planning permission to develop another deep mine south of Whitby to exploit polyhalite from the same sequence and the development is now going ahead.

2. SEISMIC PROFILES

P.R. Wood

Reflection seismology has been in use for about one hundred years and is the main tool for sub-surface mapping in the hydrocarbon and mineral industries. It requires a great deal of complex and expensive equipment and so has mainly been the preserve of oil and gas exploration and production companies since its inception. Thanks to U.K. legislation that requires the release of data after a certain period and a public service organisation, the United Kingdom Onshore Geophysical Library (UKOGL), most of the onshore seismic data in the U.K. is available online at www.ukogl.org.uk.

Reflection seismology uses differences in density and velocity (the speed of acoustic waves) of rocks to delineate different strata. A sound source is activated at the surface and acoustic waves travel through the subsurface. Where there are density and velocity variations, part of the energy is reflected and returns to the surface. The rest travels on and is in turn part reflected at increasingly deeper layers. The returning energy is picked up by sensitive detectors, 'geophones' on land, 'hydrophones' at sea, and recorded by a mobile computer system. Further computers process the data into two-dimensional profiles (seismic sections), or three-dimensional models of the Earth. These can be displayed on workstations or in 3D in virtual reality (VR) systems.

The ability to translate the results of seismic surveys into geological mapping ('interpretation') is highly variable. Offshore, surface conditions are usually reasonably constant and results are good, though they can be influenced by complex geology. Onshore, surface variations can dictate the difference between acceptable and unusable data. In NE Yorkshire, however, the data quality is among the best of that encountered onshore in the U.K. Also, much of the data on the UKOGL database come from an exploration period in the 1980s during an attempt to extend the Southern North Sea gas province onshore. At that time, the previous dynamite energy source was replaced by the Vibroseis truck-based sound source which gave improved quality.

In this section, examples are taken from the UKOGL database and some also use reprocessed seismic data from a mineral exploration project conducted by Sirius Minerals. Both UKOGL and Sirius Minerals are thanked for use of their data. A number of caveats must, however, be noted in our use of these examples. The first is that the seismic method measures time, i.e. the time it takes for the sound energy to travel into the Earth and be reflected back. To make a geological model, this has to be converted to depth, therefore we rely on calibrations at boreholes. The seismic profiles here are displayed in 'two way time', and borehole calibrations are shown where available. Most of the profiles (from the 1980s) are two-dimensional and cannot correct for three-dimensional variations, such as within the complex geology seen in the Flamborough area. In some cases there is limited (interpretable) seismic coverage or insufficient borehole sampling in complex areas to give high confidence, so the interpretation may be speculative (indicated by dashed lines). Nevertheless, use of the seismic data now available in the public domain does add a significant further dimension to our knowledge of the geology of the Yorkshire coast. The seismic sections discussed here illustrate structures from Whitby southwards to the Flamborough Fault Zone, some of which are exposed at localities described in this guide. South of the fault zone, the coastal sections only expose Quaternary deposits but the underlying Mesozoic sediments are much less disturbed.

2 Seismic profiles

2.1 Map of seismic coverage in NE Yorkshire (Fig. 7)

The base map is extracted from the UKOGL website and shows the locations in green of 2D seismic profiles available for display. The dates of seismic acquisition range from the mid 1960s to the mid 1990s. In general, because of technology improvements, only seismic profiles or 'lines' acquired after about 1980 are suitable for interpretation, that is, making a geological model from the profiles. Locations of 3D surveys are shown in grey, however, the full data sets are not available on the UKOGL web-site.

The magenta colour shows the track of a composite profile constructed by UKOGL from a number of different seismic lines. This is shown in Fig. 8. It gives a good overview of the subsurface features that can be seen on seismic from Whitby in the north to the south of Flamborough and Bridlington. The complete regional profile on the database continues south almost to the Humber.

Well locations are shown along the regional profile. Where a well velocity survey has established a relationship between acoustic travel time and depth, it is possible to calibrate the borehole stratigraphy with 'events' or 'horizons' observed on the seismic cross sections.

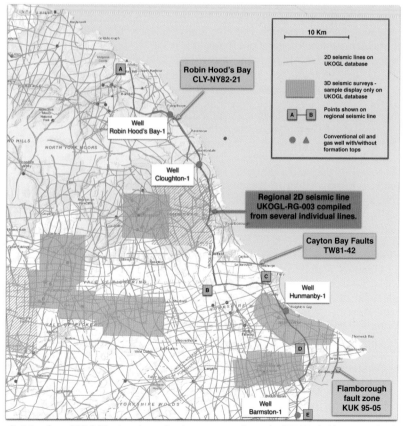

Figure 7. Map from UKOGL showing seismic coverage along the NE Yorkshire coast, including locations of seismic lines shown in this guide.

2 Seismic profiles

2.2 Regional seismic profile Whitby to Barmston (Fig. 8)

It should be noted that features interpreted onto the profile are indicative and not the result of a full workstation-based study. There are probably many more faults than those indicated and, as previously remarked, 3D seismic is necessary to map them properly. In some cases, the seismic data are too sparse or poor quality to be able to provide reliable correlations across fault blocks, especially where the structural picture is complex.

At the northern end of the profile, strata in the subsurface are relatively undisturbed, with the shallow anticline of the east end of the Cleveland inversion apparent as the Robin Hood's Bay Dome (see also Fig. 11). With the seismic acquisition parameters used, there is only a sparse image of the near surface Middle and Lower Jurassic formations that crop out on the coast. The Redcar Mudstone Formation shows little acoustic variation, so the first good seismic marker is the Rhaetic (orange) of the Upper Triassic. This starts a series of seismic reflections that is typical of much of the region until the complex fault systems of the Vale of Pickering are encountered (from point B). There is a series of near parallel reflections in the Triassic until the Bunter or Sherwood Sandstone (yellow), then a further 'transparent' section until the top of the Permian (magenta). The Permian is here represented by the Zechstein Group. The upper part is a highly reflective set of cycles between carbonates and evaporites (Tucker, 1991). The lower part is the thicker and less reflective Z2 Formation, mainly evaporites but also containing the 'polyhalite' that has been the subject of recent mineral exploration. Below this is the Variscan Unconformity with hints of complex faulting and folding in the underlying Carboniferous occasionally present.

In fact, the distinctive Zechstein seismic character can be followed on seismic data over a large part of NE Yorkshire and the event picked reliably even when others are not so clear.

Moving further south, this seismic sequence is verified at the Cloughton well before the profile moves into the complex faulting of the W-E trending system in the Vale of Pickering at point B (see Fig. 3). Because of the sparse data coverage here, the profile bends to follow a W-E track before trending south again at point C. The fault shown there is possibly the southerly part of the Red Cliff Fault in the Peak Trough (see Figs 3, 10, 11), with additional complexity added as it intersects the Pickering system.

Although it is tempting to carry the Corallian (light blue) marker straight across the faults at B and C, following other seismic sections around the regional profile shows that this is indeed a trough (although the correlations are somewhat ambiguous). This is backed up by surface evidence of the throw of the faults.

At Hunmanby the seismic can again be correlated with the stratigraphy until it moves into the complex Flamborough Fault Zone, the eastern expression of the Howardian-Flamborough system (see Fig. 3). The complexity is demonstrated in the cliffs at Flamborough Head (Itinerary 13), but the deformation and faulting seen there is just the surface expression of deep-seated and probably several phases of tectonism. Once again, apart from the nearly ubiquitous Zechstein, seismic correlation here is very difficult, especially as there are insufficient good quality seismic lines for this situation. South of Flamborough there is less disturbance, though there is a small graben between here and another calibration point at Barmston-1.

2 Seismic profiles

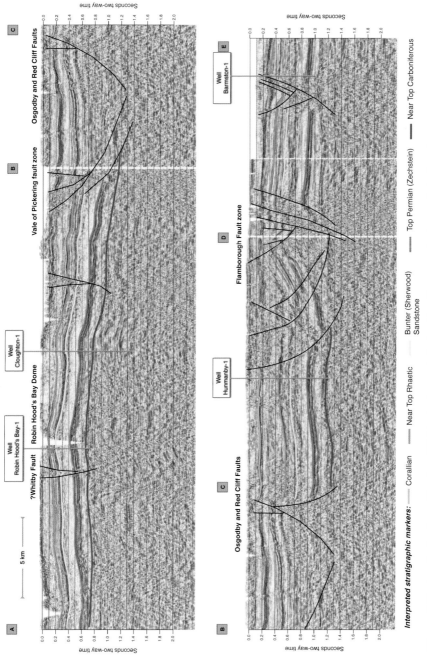

Figure 8. Regional seismic profile: from Whitby in the north (point A) to Barmston in the south (point E).

2 Seismic profiles

2.3 The Peak Trough (Figs 9–10)

The Peak Trough is a small graben system that runs sub-parallel to the coast and influences the geology from Ravenscar to south of Scarborough (see Fig. 3). The offshore expression of the Trough was mapped by Milsom and Rawson (1989) using marine seismic data. As part of the Sirius Minerals exploration programme, the same data were merged with the land data to link offshore and onshore areas. Three merged sections with a key map are shown in Figs 9-10 to illustrate the Peak Trough approaching the shore near Robin Hood's Bay and the western fault (the 'Peak Fault') crossing the shoreline at Ravenscar (Itinerary 4).

The Peak Trough has also been tracked on seismic data south of Ravenscar to the Scarborough and Cayton Bay area (see also Fig. 12).

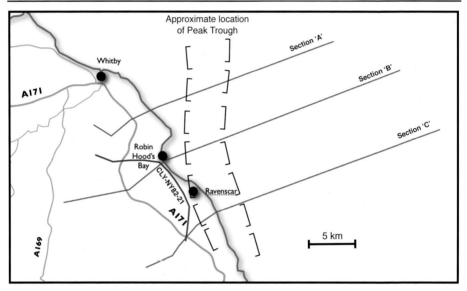

Figure 9. The Peak Trough: position of seismic lines (in red) shown in Fig. 10.

2 Seismic profiles

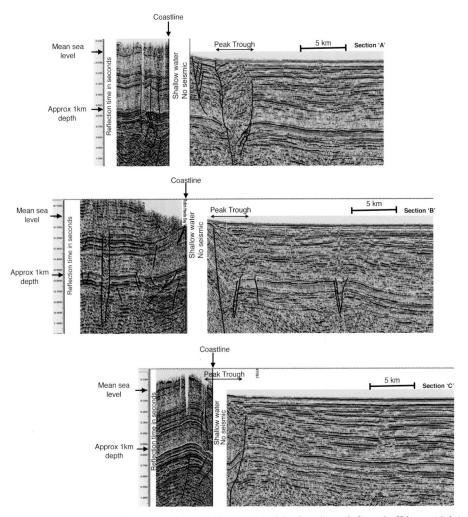

Figure 10. Seismic sections across the Peak Trough. Combined onshore (left) and offshore (right) seismic sections show the Peak Trough system approaching the shoreline at Robin Hood's Bay. Shallow-water seismic has not been acquired and is shown as a gap on cross sections. Note that colours for stratigraphic markers from this Sirius mapping project are different from those shown in the other seismic figures in this guide. The top Permian (Zechstein) is indicated by the yellow 'pick' at approximately 1 km depth on the onshore (left) part.

2 Seismic profiles

2.4 Robin Hood's Bay Dome (Fig. 11)

A further part of the Sirius Minerals programme was to reprocess seismic lines in the area around Robin Hood's Bay and further inland to get the data to a format where they could be matched together for mapping. This demonstrated that modern data processing can enhance the quality of older seismic data. In this example, seismic line CLY-NY82-21 runs from an inland location to be subparallel to Robin Hood's Bay and ends just south of Ravenscar. Most of this line was used in the regional compilation (Fig. 8) so the quality can be compared.

The small graben-like feature to the left (NW) appears to sole out in the upper Zechstein. The Variscan Unconformity (near Top Carboniferous) is delineated clearly, as is some of the underlying Carboniferous structure. There are small faults in the Kirkham Abbey carbonate above the unconformity but they do not penetrate the upper Zechstein. The Robin Hood's Bay Dome itself (Itinerary 4), a feature distinct from the overall Cleveland Anticline (Fig. 4), appears to have been, in part, formed above a salt swelling in the lower (Z2 cycle) part of the Zechstein, which here has a significant halite content.

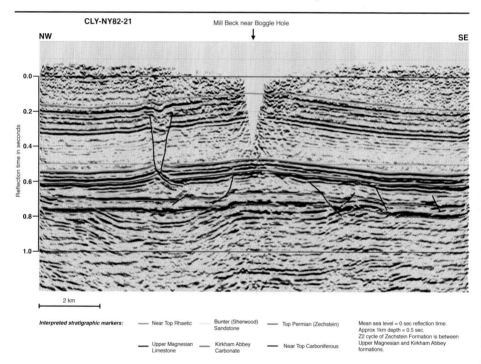

Figure 11. Reprocessed seismic line CLY-NY82-21 showing the Robin Hood's Bay Dome.

2 Seismic profiles

2.5 The Peak Trough and Cayton Bay faults (Fig. 12)

The seismic data available on UKOGL have been used to track the onshore expression of the Peak Trough south of Ravenscar. This was limited by the lack of coverage in or near the transition zone, but it appears from both seismic data and surface mapping that the western boundary of the Peak Trough is represented by an *en echelon* fault system rather than by a single continuous fault. The first E-W line south of Cayton Bay (TW81-42) shows the expression of the Peak Trough clearly, now entirely onshore. Note that the UKOGL line display has been reversed so east is to the right, matching map orientation.

The western boundary fault is defined clearly, as are two complementary bounding faults to the east, which correspond to the Red Cliff and Osgodby faults seen at either end of Cayton Bay (Itinerary 9). For clarity, the western fault is here named the Cayton Fault.

Only these major faults are shown on the seismic, but there appear to be other, minor faults in the centre of the trough.

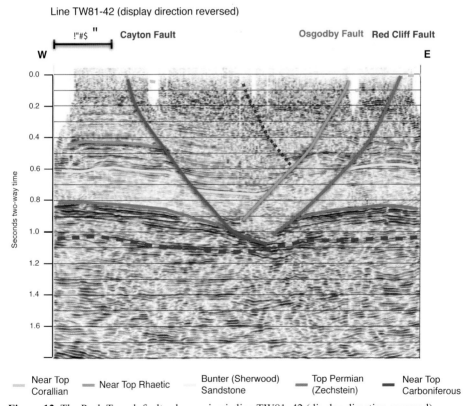

Figure 12. The Peak Trough faults along seismic line TW81–42 (display direction reversed).

2 Seismic profiles

2.6 Flamborough Fault Zone (Figs 8, 13)

The regional seismic profile (Fig. 8) shows the complexity of the Flamborough Fault Zone and the difficulty of correlating seismic reflection markers across faults, especially where there are insufficient good quality seismic lines. Line KUK95-05 (Fig. 13) is one of the few where seismic data have been acquired continuously across the shoreline, so using that line one can correlate across a gap to an E-W line south of Bridlington to calibrate with the Barmston-1 borehole.

The north end of the line appears to show a large rollover, possibly with a growth fault, or at least a low-angle one. The Zechstein is clear at either end, but the apparent 'raft' in the centre lies across the stratigraphic dip within the fault, indicating it is probably not in the plane of this cross-section, a further problem with 2D data.

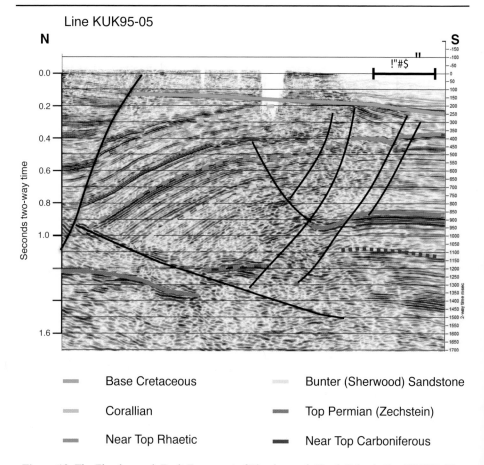

Figure 13. The Flamborough Fault Zone west of Flamborough Head. Seismic line KUK95-05.

3 Excursions: Itinerary 1

ITINERARY 1: STAITHES TO PORT MULGRAVE

P.F. Rawson

OS   1:25 000 Explorer OL 27 North York Moors Eastern area
         1:50 000 Landranger 94 Whitby & Esk Dale
GS   1:50 000 Sheet 34 Guisborough

The 3 km stretch of coastline between Staithes and Port Mulgrave provides a magnificent series of exposures of the Lower Jurassic Staithes Sandstone, Cleveland Ironstone and lower part of the Whitby Mudstone formations. The rugged splendour of the area has been much modified by man; the former working of ironstone, alum shale and jet in the area has left extensive evidence (Goldring, 2001), some of which is mentioned below. The skyline to the west of Staithes is dominated by the Boulby potash mine, which opened in 1968, is up to 1400 m deep and now has over 1000 km of tunnels. It also houses a specialist underground science laboratory designed to take advantage of the almost complete freedom from interference by natural background radiation in the rock salt layers.

The route is mainly over a rocky wave-cut platform and a falling tide is essential as the rising sea reaches the foot of the cliff in places, especially close to the harbour, by about mid-tide. Unfortunately, a major landslide in 2016 destroyed part of the path up the cliff from the derelict harbour at Port Mulgrave. At the time of publication a short, steeply sloping ladder and steps composed of vertical wood boards have been installed to replace the missing section but are very muddy in wet weather. Although the following account covers the whole section to Port Mulgrave, as did previous editions, followers of this itinerary, particularly parties, are advised to proceed no further than Brackenberry Wyke before returning to Staithes village.

Locality 1. Staithes to Penny Nab (Staithes Sandstone Formation)

Staithes is a picturesque fishing village which retains much of its original character—though many of the cottages are now second homes. It is not possible to park in the lower part of the village: from the clearly signed village car park (Fig. 14) walk down the main road, noting to the left the 30 m deep gorge cut through the Staithes Sandstone Formation by the postglacial Staithes Beck. Continue through the older part of the village to the beach within the harbour, from where the Staithes Sandstone Formation (28.6 m thick) can be viewed in the cliffs on both sides of the harbour. Howarth (1955) showed that the lower part of the formation belongs to the Davoei Zone (12.6 m thick according to Howard, 1985) and the upper part to the Margaritatus Zone (16 m thick). It consists of shallow-marine sandstones and siltstones with clearly displayed sedimentary structures. The thicker-bedded units are often cross-bedded; they include excellent examples of hummocky cross-stratification, which formed through the reworking of shallow sands by oscillating storm waves. Thinner-bedded units consist of sheets of fine sandstone fining upwards to mudstone, the base of individual sheets being erosive. They show delicately preserved parallel lamination, low-angle cross-lamination and wave-ripple lamination, but in many cases this is at least partially destroyed by bioturbation (mainly *Chondrites*), which often becomes intense in the more argillaceous upper part of each sheet.

The lowest part of the formation crops out at Cowbar Nab, to the west of the harbour, but at shore level is now largely obscured by rock armour. However, all but the lowest beds are readily accessible to the east.

3 Excursions: Itinerary 1

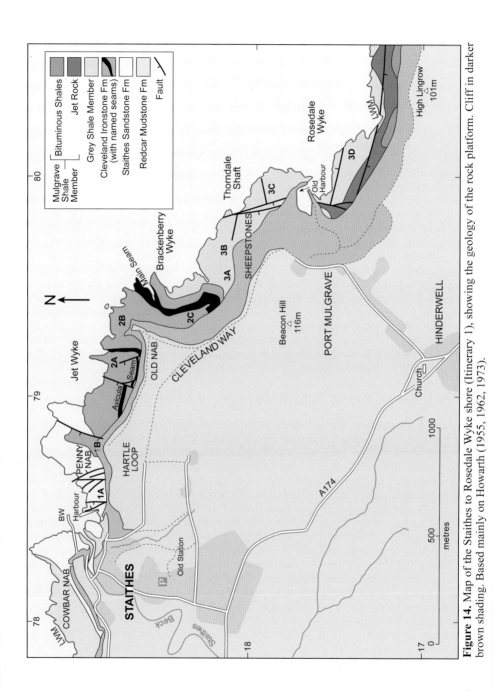

Figure 14. Map of the Staithes to Rosedale Wyke shore (Itinerary 1), showing the geology of the rock platform. Cliff in darker brown shading. Based mainly on Howarth (1955, 1962, 1973).

1A. Eastern side of Staithes Harbour. The higher part of the Staithes Sandstone Formation forms the lower part of the cliff and adjacent scars, though in the immediate vicinity of the harbour wall it is partly hidden beneath rock armour. Higher in the cliff the individual ironstone bands of the Cleveland Ironstone Formation stand out clearly. In the vertical faces at the cliff foot inside the harbour all the sedimentary features noted above can be seen. On the scars to the east of the harbour wall, ripple-marked surfaces are visible and sideritic concretions preserve large numbers of the bivalves *Protocardia truncatum*, *Oxytoma cygnipes* and *Gryphaea depressa*, and the scaphopod *Dentalium giganteum*.

Several minor faults occur here, especially in a small recess in the cliff known as Hartle Loop, about 150 m east of the harbour.

1B. Penny Nab. Eastwards towards Penny Nab the Staithes Sandstone Formation becomes increasingly argillaceous upwards until it grades into the cyclic sediments of the overlying Cleveland Ironstone Formation. The base of the latter formation (and of the Penny Nab Member) is taken at the base of the first cycle (Howard, 1985), i.e. at the base of Howarth's (1955) bed 24. This is a row of scattered siderite mudstone nodules, sometimes packed with small ammonites (*Amaltheus stokesi*), occurring round the foot of Penny Nab at the base of a sloping ledge. A few paces to the north-east of this ledge a series of parallel grooves in the shale at regular 1.2 m (4 feet) intervals marks the line of an old tramway for transporting ironstone to a shallow dock north-west of the Nab (Owen, 1985, figure 2).

Locality 2. Jet Wyke to Brackenberry Wyke (Cleveland Ironstone Formation)

In the type area of the Cleveland Hills the Cleveland Ironstone Formation contains thick ironstone seams which were formerly mined extensively. The ironstones thin and the intervening shales thicken towards the coast, where the formation is excellently exposed in Jet Wyke and round Old Nab into Brackenberry Wyke. Here, it consists of 25.3 m of shales and thin siltstones with sideritic and chamositic ironstone seams, some of which are oolitic in texture. Most seams cap coarsening-upwards cycles up to 7 m thick (Fig. 15), and individual cycles are laterally continuous over much of the basin (Rawson *et al.*, 1983; Howard, 1985). The upper part of some cycles is striped with thin fining-upwards sheets, sometimes with basal gutter marks, probably deposited under storm conditions ('tempestites'). The formation is divided into the predominantly shaly Penny Nab Member (18 m) and the more ferruginous Kettleness Member (7.3 m) (Howard, 1985).

2A. Jet Wyke. The whole of the Cleveland Ironstone Formation is accessible here, though the highest beds are better examined around Old Nab. The succession dips gently eastwards and several faults repeat parts of it, in one case bringing ironstone against ironstone in the middle of the Wyke to create a very extensive ironstone pavement on the shore. A detailed lithological sequence was given by Howarth (1955) and this has been combined with more recent sedimentological work to show the whole sequence in a log (Fig. 15). The lowest ironstone, the Avicula Seam, forms a flat ledge starting about 150 m east of Penny Nab. Its upper surface shows many specimens of *Oxytoma* (formerly *Avicula*) *cygnipes*, while the base is conglomeratic.

Further east, the 'upper striped bed' of Greensmith *et al.* (1980) is often cleanly exposed at the cliff foot immediately beneath the thin (10 cm) Raisdale Seam ironstone. This bed (about 2 m thick and forming the upper part of bed 34) consists of a series of delicately preserved layers of pale coloured, laminated siltstone fining up to darker

3 Excursions: Itinerary 1

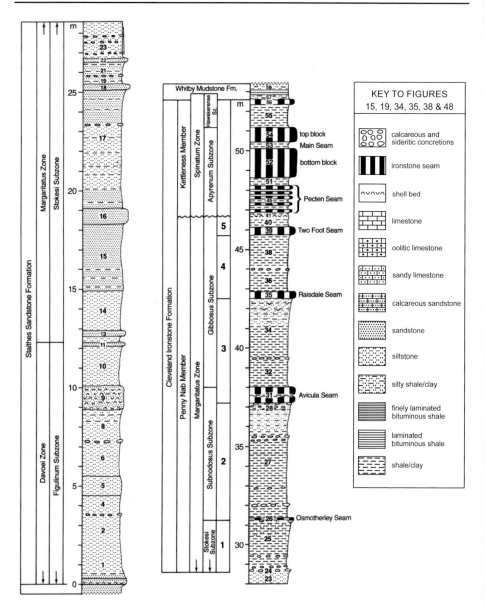

Figure 15. Lithic log of the Staithes Sandstone and Cleveland Ironstone formations at Staithes. Modified from Rawson and Wright (1996, fig. 22): based mainly on data in Howarth (1955); Howard (1985); Knox *et al.* (1991).

mudrock (Fig. 16). Each layer has an erosive base and gutters are developed at the base of some, cutting down into, and even undercutting, up to six underlying layers. The bed can be traced from the cliff foot onto the scars towards Old Nab, where the lighter, hard bases of the anastomosing gutters are seen in plan to be up to 0.5 m wide and 5 m long with an orientation almost due east-west. Similar gutters in the correlative bed at Hawsker (20 km to the south-east) are much less deep and have a finer-grained infill as if they are more distal from the shoreline, indicating currents from the west. The sequence at both localities suggests deposition under storm surge conditions.

Along the eastern side of the Wyke the Pecten Seam is seen in vertical section to be represented by five thin layers of ironstone separated by thin shale partings. The base of the seam is taken as the base of the Kettleness Member. Above this level the former mining of the overlying Main Seam is evident higher in the cliffs, marked by a series of adits into which the overlying shales have collapsed.

2B. Old Nab. The regular blocks at the tip of this prominent headland are again the result of ironstone mining; pillars of Main Seam ironstone left as roof supports and the intervening bords (tunnels) have since been unroofed by marine erosion. On the east side of the Nab, shale backfill can be seen filling an adit running into the cliff. In the immediate vicinity of the Nab and on the intertidal scars in Brackenberry Wyke the Main Seam bedding surfaces show extensive networks of *Rhizocorallium* (crustacean burrows), most of them showing scratch marks made by the crustaceans' claws.

Figure 16. The 'Upper Striped Bed' (bed 34 upper) in Jet Wyke. This consists of thin fining-upward sheets with erosive bases and well-developed gutters. The base of the overlying Raisdale Seam is conglomeratic.

2C. West side of Brackenberry Wyke (Fig. 17). Here, the higher part of the Cleveland Formation is gently arched so that the lowest beds are seen about halfway along the cliff. Mine adits are again visible and ironstone was also quarried from the shore. There are good exposures of bedding surfaces of the Pecten and Main Seams, the former crowded with *Pseudopecten aequivalvis* and the latter riddled with *Rhizocorallium* (Fig. 18). Towards the head of the Wyke the ironstone nodules of bed 56 form the last ironstone platform at the cliff foot. They contain body chambers of *Pleuroceras hawskerensis* plus numerous bivalves (*Pleuromya costata* and *Gresslya*) still preserved in burrowing position.

Immediately above bed 56, and clearly seen low in the cliff face, a sharp facies change occurs, marking the base of the Grey Shale Member of the Whitby Mudstone Formation and the basal Toarcian transgression. About 0.45 m above the base is a distinctive 0.2 m thick finely laminated shale with lenticles of jet, while the overlying sediments consist of grey micaceous mudstones with concretionary horizons.

Both the ironstone of bed 56 and the overlying Grey Shale Member, which includes six bands of red-weathering sideritic concretions, can be traced from the cliff foot across the rocky platform at the south end of the Wyke. Here, the shingle on the shore above consists largely of nodules derived from various levels in the Whitby Mudstone Formation and is a good source of ammonites. In the cliff face about 2 m above the shore there is a mine adit in the Jet Rock, which forms the lower part of the cliff face here.

It is from this locality that users of this guide are advised to return to Staithes to avoid the difficult path up the cliff at Port Mulgrave. For those who wish to continue and take that path, note that the walk around the southern corner of Brackenberry Wyke and past Thorndale Shaft towards Port Mulgrave harbour can be difficult because of boulders and weed-covered rock scars.

Locality 3. Brackenberry Wyke to Port Mulgrave (Whitby Mudstone Formation)

3A. The Sheep Stones. On the southern corner of Brackenberry Wyke the shore is strewn with well-weathered blocks of sandstone that have fallen from the Saltwick Formation at the top of the cliff. On walking round this area note that the more easterly blocks in particular are often firmly embedded on shale stacks up to half-a-metre high. These are the Sheep Stones, which have been interpreted to represent an ancient fall that took place when the rocky intertidal platform and mean sea level were slightly higher, probably during the Ipswichian (= penultimate) interglacial period (Agar, 1960). Part of this area has been covered by massive cliff falls which extend round to the adjacent part of Thorndale Shaft, bringing down material from the Alum Shale Member of the Whitby Mudstone Formation and blocks of sandstone from the Saltwick Formation.

3B. Thorndale Shaft. The shales on the shore to the south-east of the rock falls belong to the upper part of the Grey Shale Member. The characteristic ammonite *Dactylioceras tenuicostatum* occurs mainly in nodules which have been extensively collected, so that some of the principal nodule beds are represented by lines of hollows in the scars. The highest beds contain *D. semicelatum*, joined in the top 1.8 m by *Tiltoniceras antiquum* (see section in Howarth, 1973). Towards the cliff foot the Jet Rock, the lowest informal division of the Mulgrave Shale Member (Fig. 19), crops out and forms the whole of the accessible lower part of the cliff between here and Port Mulgrave harbour. The Jet Rock is a finely laminated, dark-coloured, richly pyritic, bituminous shale with numerous concretions either in beds or randomly scattered. These are usually pyrite-skinned, very hard and splintery and therefore very dangerous to hammer. There are four main marker beds here,

3 Excursions: Itinerary 1

Figure 17. The western side of Brackenberry Wyke to Old Nab. The rock scars and lowest part of the cliff are formed by the Kettleness Member, the remainder of the cliff by the Grey Shales Member. Towards Old Nab, former mine workings of the Main Seam of the Kettleness Member are visible, now unroofed by cliff recession.

Figure 18. *Rhizocorallium* in the Main Seam, Kettleness Member, Brackenberry Wyke.

the Cannon Ball Doggers (which form the base of the member), Whalestones, Curling Stones and Top Jet Dogger (Fig. 19). Near to the harbour the horizon of the Cannon Ball Doggers is marked by a shale scar adjacent to the cliff foot, with numerous hollows where the doggers have been removed. The remaining marker beds are seen in the cliff face, where the Top Jet Dogger forms a thin (25 cm), continuous impure limestone band about 5 m above shore level, with an occasional flat, large concretion ('Millstone') in the upper part. This marks the top of the Jet Rock. Thus jet should be searched for in the lower part of the cliff. It occurs as long, compressed seams, lustrous black in cross section, and is most easily found after winter storms when cliff falls have occurred. Later in the year the former position of seams is marked by small excavations in the cliff face! The specialised bivalve *Pseudomytiloides dubius* is common here, and flattened, pyritised ammonites are visible on the scars, often in current-swept accumulations.

A mine adit formerly visible in the cliff face has now been hidden by a landslip.

3C. Port Mulgrave. Port Mulgrave harbour was built in 1856–7 to ship ironstone worked from the shore and adjacent mines to the iron works of Tyneside. The bricked-up entrance to the main mine roadway (Seaton Drift) is visible in the cliff. Plans of the mining area and detailed information on the history of the mining are given by Owen (1985).

3D. Rosedale Wyke. The Grey Shale and Mulgrave Shale members form the rocky platform here (Fig. 14) and show the same marker beds seen in the previous exposure. The Jet Rock was formerly quarried from the shore and the main marker beds are readily traced, though partly displaced by small faults running slightly obliquely to the bedding. The Whalestones form distinctive small stacks while the Curling Stones and closely associated Upper Pseudovertebrae are also prominent a little higher in the sequence. Above are the Millstones, lenticular doggers of limestone up to 3 m in diameter set in the upper surface of the Top Jet Dogger; these occur close to the cliff foot in the middle of the Wyke and are often covered by sand and fallen shale.

Return from here to the adjacent harbour where the path and ladder up the cliff leads to a cliff-top footpath (the Cleveland Way) back to Staithes.

3 Excursions: Itinerary 1

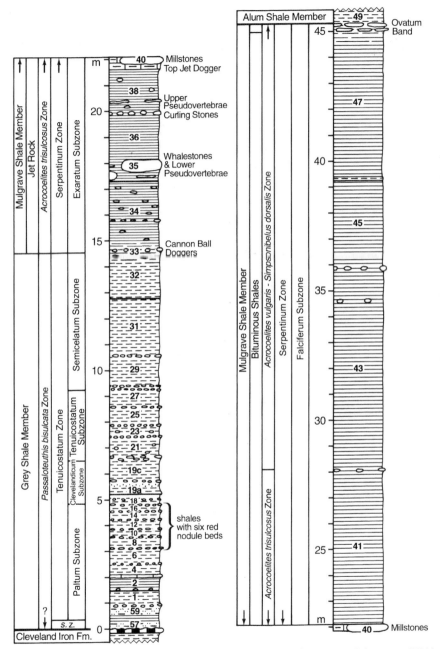

Figure 19. Lithic log of the Grey Shale and Mulgrave Shale members: Port Mulgrave to Whitby. For lithological key see Fig. 15. Modified from Rawson and Wright (1996, fig. 23); based mainly on data in Howarth (1962, 1992).

ITINERARY 2: RUNSWICK BAY

P.F. Rawson

OS   1:25 000 Explorer OL 27 North York Moors Eastern area
         1:50 000 Landranger 94 Whitby & Esk Dale
GS   1:50 000 Sheet 34 Guisborough

From Whitby head north along the A174, turning off to the right at NZ 800 150 onto a minor road signposted for Runswick Bay. The road terminates at an oblique T-junction: turn right, then almost immediately right again (ignore the signpost for a car park straight ahead, which is the cliff-top one). Follow the road downhill to a public (pay) car park on the right, positioned between private ones.

    The highest beds of the Cleveland Ironstone Formation are the oldest rocks seen in this itinerary, and much of the sequence here consists of dark shales of the Grey Shale and Mulgrave Shale members of the Whitby Mudstone Formation. The itinerary falls into two parts: exposures on flat scars to the NNE of the village, and on scars and at the cliff-foot ESE of the village (Fig. 20). It is advisable to examine the northern exposure first, while the tide is falling.

Locality 1. Topman Steel (Mulgrave Shale Member)

From the car park head for the shore by the village, walking northwards past the lifeboat station and skirting around the seaward side of the newly installed (2018) rock armour which abuts the earlier sea-defence wall. Although much of the shore here is strewn with weed-covered boulders the area immediately in front of the rock armour is reasonably clear. Adjacent to the northern end of the rock armour grey shales belonging to the lower part of the Bituminous Shales (Mulgrave Shale Member) are visible among the boulders on the shore and include a layer of large oval concretions which form bed 42 of Howarth's (1962) section (Fig. 19). The concretions are pyrite-skinned and fragments of pyritised *Pseudomytiloides* project from them. From here the underlying shales of bed 41 form a weed-covered platform traversed by a narrow weed-free path for about 100 m where it reaches a much more extensive flat-lying area, Topman Steel. Here there is an extensive exposure of part of the Mulgrave Shale Member, showing the Jet Rock and the lowest part of the Bituminous Shales. The beds strike obliquely across the shore and dip gently south-westward so that one is working down section northwards towards low water. The first well-exposed shales encountered represent the lowest part of bed 41, which forms the basal bed of the Bituminous Shales. Belemnites and flattened, often pyritised, ammonites (*Harpoceras falciferum* and occasional *Hildaites murleyi*) occur. Both here and in the underlying beds the ammonites sometimes occur in small, current-swept accumulations.

    The immediately underlying Millstones (bed 40) are beautifully exposed here. As at Port Mulgrave these are almost perfectly circular, lenticular, calcareous concretions up to about 3 m in diameter but with a maximum thickness of only about 0.3 m. They sit in the upper part of a hard laminated, argillaceous limestone, the Top Jet Dogger (bed 39) which forms a small scar across the shore (Fig. 21). The shales visible immediately below are very finely laminated and some laminae show clear evidence of erosion at the base. They include two closely spaced concretionary levels, the Upper Pseudovertebrae and the Curling Stones. They are quite fossiliferous, again with flattened, often pyritised, ammonites (*Harpoceras*, *Cleviceras* and *Dactylioceras*) and solid belemnite guards. Further down

3 Excursions: Itinerary 2

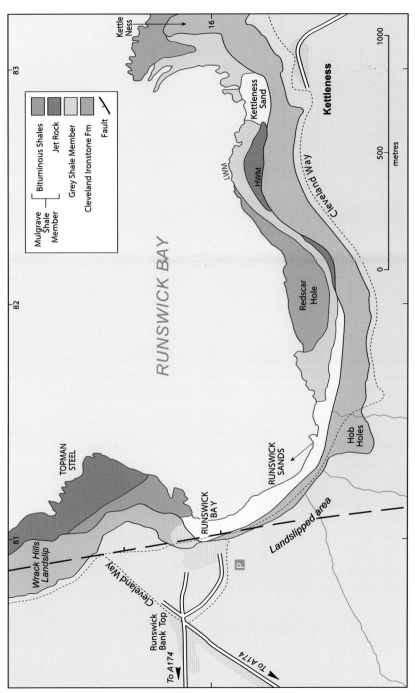

Figure 20. Map of Runswick Bay to Kettle Ness point (Itinerary 2), showing the geology of the rock platform. Cliff in darker brown shading. Based on data in Howarth (1962), BGS (1998a) and P. F. Rawson's field observations.

3 Excursions: Itinerary 2

Figure 21. The Millstones and Top Jet Dogger, Topman Steel, Runswick Bay. The Millstones are the discoidal masses on the left-centre of the photo, the Top Jet Dogger the pale band on the right-hand side.

the sequence the laminated shales become partially hidden by boulders, but towards low-water mark the Whalestones can be seen. These are large (up to 2 m long and 0.5 m thick) calcareous doggers often sitting on eroded pillars of the underlying shale.

In the cliff adjacent to this exposure, various marker beds in the Bituminous Shales, up to the distinctive double nodule band forming the Ovatum Band at the top of these shales, can be seen in the vertical face. The higher part of the succession (Alum Shale Member) is very weathered but at the top of the cliff are some well-developed channel sandstones at the base of the Saltwick Formation, readily visible from the shore. To the right (north) of this high cliff is a landslipped area known as Wrack Hills, on which the remains of two ironstone calcining kilns are visible; these are relics of the former Albert Iron and Cement Works (Owen, 1988; Goldring, 2001). The change in topography in the cliff here marks the position of a N-S fault running through the western side of the bay. On the shore beneath this area, fallen blocks of Saltwick Formation sandstones include plant remains and an occasional dinosaur footprint.

Locality 2. Redscar Hole to Runswick village (Cleveland Ironstone Formation to Mulgrave Shale Member)

Return to the village and head south-eastward across the sandy shore towards the first headland, walking onto an extensive rock pavement to a slight embayment on part of the shore called 'Redscar Hole' [NZ 822 156] on large-scale maps (Fig. 20). From here one can work back towards the village, following the sequence upward.

The eastern half of the rock pavement exposes a distinctive bed of red, nodular ironstone with a very irregular upper surface. This forms the top of the Cleveland Ironstone Formation, and is bed 24 of Howarth's (1955) Kettleness section. It is the same as bed 56 of the Staithes section and as at Staithes it yields *Pleuroceras hawskerense* and burrowing bivalves preserved in life position. This bed extends almost to the cliff foot at Redscar Hole and about 0.3 m above, on a slightly higher ledge beneath the cliff face, is the distinctive thin, finely laminated bituminous shale (Bed 26) which is also seen at Staithes (Bed 58) and lies just above the base of the Grey Shale Member. About 2 m hgher, visible in the lower cliff face, are the six red nodule bands of Howarth's (1962) section. Moving westward, these beds dip down onto the shore to form the western half of the rock platform, where much of the Grey Shale Member is weed covered, though the red nodule bands are sometimes visible and yield occasional *Dactylioceras tenuicostatum* and *D. semicelatum*.

Continuing westward, higher levels appear in the cliff, as the Grey Shale Member passes upward into the Jet Rock (Mulgrave Shale Member). The last part of the cliff exposing solid rock lies between two small ravines formed by Claymoor and Calais becks. Here the Jet Rock is readily accessible and shows clearly the very fine-scale lamination characteristic of these beds. The remains of several mine adits are visible ('Hob Holes' on OS maps).

Westward from here to the village glacial tills appear down to sea level.

Note: A partial alternative to this itinerary is to ignore locality 1 (Topman Steel) and instead walk eastward from Redscar Hole round a small promontory for about 400 m over boulders and rubble (hugging the cliff foot avoids some of the boulders) to reach Kettleness Sand, where excellent exposures of the Mulgrave Shale, Grey Shale and upper Cleveland Ironstone formations extend to Kettle Ness [NZ 832 162] (see Howarth, 1955, 1962, 1973 for details of the sections). Tidal considerations mean that it is impossible to combine this alternative itinerary with the Topman Steel exposures. Note also that there is no way up the cliffs from Kettleness Sand or Kettle Ness: one has to return to Runswick Bay.

ITINERARY 3: SALTWICK BAY TO WHITBY

P.F. Rawson and J.K. Wright

OS   1:25 000 Explorer OL 27 North York Moors Eastern area
        1:50 000 Landranger 94 Whitby & Esk Dale
GS   1:50 000 Sheet 35/44 Whitby and Scalby

This short excursion examines the Whitby Mudstone Formation and part of the Ravenscar Group. It follows the rock platform at the base of the cliffs from Whitby to Saltwick Bay, ascending the footpath in the bay and returning to Whitby via the cliff-top path (Fig. 22). The going is generally easy, though the shales can be very slippery. **This is strictly a low tide itinerary. The headland east of Whitby East Pier and the end of Saltwick Nab are passable for less than two hours either side of low tide, so the itinerary should not be attempted on a rising tide.** Parking is available on the Endeavour Wharf car park on the opposite side of the harbour, though the easier option is to drive to the Abbey car park and descend to the shore via the Abbey steps: the car park is then a short walk away from Saltwick Bay at the end of the excursion.

Locality 1. Whitby East Pier (Whitby Fault)

From the foot of the Abbey Steps walk along Henrietta Street and follow the path down onto the East Pier to view the contrasting successions in East and West cliffs and adjacent shore. To the east, the rock platform and lower cliff expose the top 12 m of the Alum Shale Member. Above is the Dogger Formation, here a ferruginous sandstone only 0.75 m thick. The overlying fine sandstones, siltstones and carbonaceous clays of the Saltwick Formation (*c.* 31 m) pass up into the marine Eller Beck Formation (*c.* 6 m), which is capped by glacial till (Fig. 23).

By contrast, the West Cliff and Khyber Pass expose a solely non-marine sequence of channel sandstones stacked one above the other. This striking difference suggests the occurrence of a NNE/SSW fault with a westerly downthrow running along the deep channel of the harbour. The West Cliff succession was provisionally assigned to the 'Middle Estuarine Series' (= Cloughton Formation) by Fox-Strangways and Barrow (1915) in the middle of the Ravenscar Group. However, subsequent detailed mapping, particularly of the Dogger Formation (Hemingway, 1958; British Geological Survey, 1998a), borehole evidence within the harbour, and spore analysis (Harris, 1953) shows that the West Cliff succession is part of the Saltwick Formation. This indicates that the fault (which is not exposed) downthrows no more than 12 m to the west. Osborne (1998, pp. 266–286) gives a detailed account of the history of research into the Whitby Fault. Alexander (1986) has proposed that it was active in Mid Jurassic times and that, on the downthrow side to the west, was a persistent, low-lying area repeatedly occupied by river channels. This would explain the contrast within the Saltwick Formation between the succession of infilled channels visible to the west and the largely level-bedded sediments to the east.

Locality 2. Whitby East Cliff to Saltwick Nab (Whitby Mudstone, Dogger and Saltwick formations)

2A. Cliffs and shore beneath the Abbey. From East Pier walk down a concrete slope to the shore. Here the adjacent cliffs are eroded into four small embayments ('bights'). The

3 Excursions: Itinerary 3

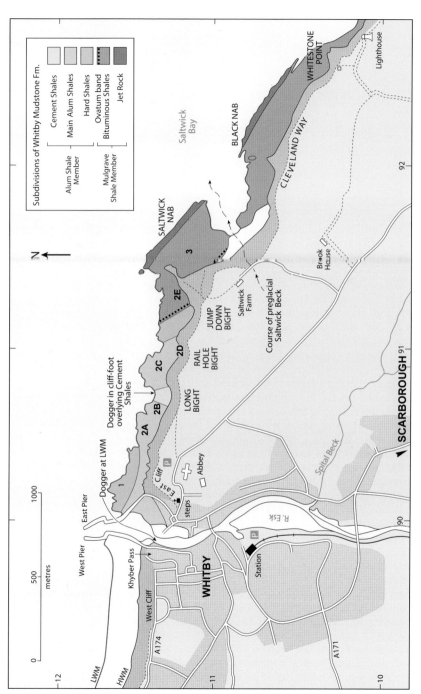

Figure 22. Map of Saltwick Bay to East Cliff, Whitby (Itinerary 3), showing the geology of the rock platform. Cliff in darker brown shading. Based on data in Howarth (1962).

3 Excursions: Itinerary 3

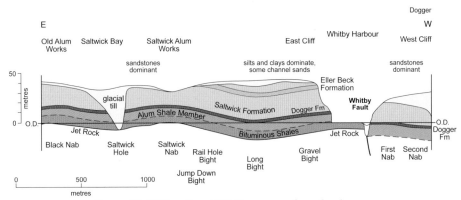

Figure 23. Cliff section at Whitby, as seen from the shore.

vicinity of the cliff foot is inaccessible along this part of East Cliff, being protected by rock armour, mainly large blocks of gneiss, but on the shore there are extensive exposures of the Alum Shale Member, particularly in the intertidal zone. This member is informally divided into three units. The lowest, the Hard Shales, crops out on the lower part of the shore, while the overlying Main Alum Shales are exposed on the higher part. Within the latter unit, a bed of scattered, red-weathering sideritic doggers, often with a more calcareous, greyish core, forms a distinctive marker along the upper part of the scar. This is bed 52 of Howarth's (1962) section. The Main Alum Shales are very fossiliferous; there are abundant specimens of the thick-shelled, shallow-burrowing *Dacryomya ovum*, which occur mainly crushed, some uncrushed, and occasionally in sufficient abundance to form thin limestones. Occasional *Dacryomya* are preserved in life position, as also are *Pleuromya* sp. Belemnites and ammonites (*Dactylioceras* and *Hildoceras*) are also common, the latter usually with only the outer whorl (body chamber) well preserved, the inner whorls being crushed, sometimes within a calcareous mudstone concretion.

The lower part of the cliff face visible above the rock armour exposes the highest unit of the Alum Shale Member, the Cement Shales, which obtain their name from the abundant calcareous nodules that can be seen forming several discrete beds in the shale. About 12 m above shore level the first hard rock band is the Dogger Formation, which is overlain by the non-marine beds of the Saltwick Formation.

The last small headland in this stretch of protected cliff is the critical point for access from or to Whitby. Here the rock armour extends furthest seaward, so **the sea is only free of the area for little more than 1½ hours either side of low water and great care must be taken not to get cut off.**

2B. Vicinity of the old rockfall to Long Bight. From this headland to a small embayment known as Long Bight is a 350 m stretch of almost straight cliff. From the foot of the headland bed 52 swings out towards low water, forming a clear feature on the almost flat scars formed by the Main Alum Shales. About 200 m along from the headland, the cliff protrudes slightly and here there is a large rock fall, part dating from about 1912. Approaching the rock fall, note that the Dogger Formation in the cliff thins and is eventually cut out due to erosion at the base of a broad channel in the overlying Saltwick Formation. The western side of the channel is filled with gently dipping sandstone/siltstone alternations. Eastwards, a large incursion of sand into the channel system is marked by thick, easterly dipping cross-sets (epsilon cross-bedding). These were laid down on the inside of a bend, as a point bar, as the channel migrated eastwards. On the eastern side of the original rock fall the sediments are partially hidden by more recently fallen boulders, but eastward of these into Long Bight is a shallow, shale-filled channel (Fig. 24) which cuts down into the eastern, thinner part of the main channel.

The eastern side of the shale-filled channel cuts into up to 2.5 m of thinly bedded, fine-grained, pale grey sandstones with occasional concretions of sideritic mudstone. These initially dip eastward but along the cliff face became more horizontally bedded and are seen to rest on the Dogger Formation which here forms a shelf at the foot of the cliff dipping gently to the west. This reflects the development of a shallow syncline in Long Bite that brings both the Dogger Formation and the Cement Shales down to shore level. Here the Dogger is a tough, very pebbly sideritic sandstone, 0.4 m thick. Abundant U-shaped tubes allied to *Arenicolites* or *Diplocraterion* descend from its base into the Cement Shales. The latter form a scar on the adjacent shore and yield *Catacoeloceras* and, less commonly, *Hildoceras bifrons* and *Hildoceras hildense*. There are also abundant belemnites, mainly *Salpingoteuthis*.

The fine sandstones immediately above the Dogger Formation yield a flora including *Coniopteris*, *Williamsonia*, *Baieria* and *Czechanowskia*, and are known as the Whitby Plant Bed. This is evidently an offbank deposit laid down under quiet conditions marginal to the main, sand-filled channel.

About 3 m above the Whitby Plant Bed is an up to 0.5 m thick pale cream to orange-weathering, fine sandstone riddled with vertical rootlets forming a very distinctive marker in the cliff. Loose blocks of it are common at the cliff foot, as are blocks of a fine-grained

Figure 24. Shale-filled channel in the Saltwick Formation, East Cliff, Whitby.

sandstone, fallen from a horizon about 5.5 m above the Dogger Formation and containing paired valves of *Unio kendalli*, sometimes gaping, which may equate with the freshwater shell-bed originally described at Saltwick (Jackson, 1911).

2C. Headland between Rail Hole Bight and Long Bight. On the headland, the cliff is free of debris down to high water mark and a buttress of Dogger projects from the base of the cliff. It was 10 m west of the northernmost tip of the headland that Whyte and Romano (1993, fig. 4) discovered a large, fallen, rotated block of Saltwick Formation sandstone displaying on its undersurface raised, infilled moulds of footprints showing the broadly triangular 3-toed hindfeet and small, crescentic forefeet of an early form of stegosaur (Whyte & Romano, 2001; Romano & Whyte, 2015). This block lay within the range of high spring tides and is no longer available for study. However, blocks of Saltwick Formation sandstone regularly fall down the cliffs here and may reveal a variety of dinosaur footprints: visitors are asked to keep a close watch for new blocks displaying footprints and report them to Whitby Museum.

2D. Rail Hole Bight. The Main Alum Shales are well exposed on the shore, and are very fossiliferous again. A distinctive sideritic ironstone band (bed 52) runs from low water mark north of the bight south-eastwards to the foot of the headland on the eastern side of Rail Hole Bight. As the sequence is traced upwards from this bed within the bight, the last true *Dactylioceras* (*D. athleticum*) are abruptly replaced by *Peronoceras*, a dactylioceratid with spinose, fibulate ribs.

In the cliff face, the Dogger Formation climbs the cliff eastwards with a largely channel-free Saltwick Formation succession above it.

2E. Jump Down Bight and the western side of Saltwick Nab. Jump Down Bight is the last embayment before Saltwick Nab. The Hard Shales form the foreshore in the eastern part of the bight, the top being marked by a continuous 0.2 m thick sideritic mudstone (bed 50) crossing the shore obliquely about 40 m north-east of the headland between Rail Hole and Jump Down Bights and running toward the middle of the latter. These shales pass down into the Bituminous Shales (Mulgrave Shale Member) towards the north-east. The boundary of the two lies at the top of the Ovatum Band, the highest of the three informal units forming the Mulgrave Shale Member. The Ovatum Band (bed 48) is visible in the base of the cliff in the southernmost corner of the bight and from there can be traced north-westward across the foreshore. It is a distinctive 25 cm thick double band of pyrite-skinned, discoidal sideritic concretions that yield rare *Ovaticeras ovatum*. Some of the concretions are set in larger sheets of reddish-weathering sideritic limestone and occasional large masses of belemnite-limestone up to 7 cm thick also occur.

The passage from the Mulgrave Shale Member to the Alum Shale Member is marked by a change from finely laminated to non-laminated shales. This upward change in lithology reflects a return to better-oxygenated conditions on the sea floor following the Toarcian OAE and is marked by the gradual return of mud-burrowing bivalves. Thus *Dacryomya ovum* first becomes common in the lowest part of the Main Alum Shales, in the western part of Jump Down Bight. Belemnites are abundant and well preserved here.

Northward from Jump Down Bight the Bituminous Shales are exposed in shore and cliffs to the end of Saltwick Nab. For tidal reasons it is safer to examine these beds round the eastern side of the Nab in Saltwick Bay, either by walking round the Nab at low tide, or over the notch near its neck—but note that the shales at the latter are steep and very slippery; the best footing is on some of the more barnacle-covered shale 'steps'.

Locality 3 Saltwick Bay (Mulgrave Shale Member)

The shales that crop out so extensively in Saltwick Bay belong to the Mulgrave Shale Member, mainly to the Bituminous Shales (Fig. 22). Howarth (1962) has published a detailed bed-by-bed account of the sequence here, which is summarised in Fig. 19.

Near low water is a well-marked reef made of a tough, impure limestone (the Top Jet Dogger) that marks the top of the lowest division of the Mulgrave Shale Member, the Jet Rock. As at Runswick Bay, this limestone is characterised by large (up to 4.2 m in diameter) discoidal concretions, the Millstones, set in the top surface of the limestone. Note the very well-marked jointing of this bed and the adjacent shales. Below it, the Jet Rock was formerly worked extensively for jet here, even though it is accessible only at low water spring tides.

The lowest part of the Bituminous Shales is usually weed covered, but exposures become much clearer as one works up the sequence from just below the foot of Saltwick Nab across the shore towards the cliffs. The extensive exposures across the intertidal zone show finely laminated shales with four clear nodular marker bands. The first is Howarth's (1962) bed 42, a row of scattered oval doggers with pyritic skins that skirt the seaward foot of Saltwick Nab. These distinctive doggers include pyritised *Pseudomytiloides* on their upper surfaces and are seen also at Runswick Bay. About halfway up the eastern side of the nab is bed 44, a row of scattered doggers with pyrite aggregations while near the top of the nab is bed 46, a red-weathering sideritic mudstone (Fig. 25). All three bands can be traced running east-south-east across the scars within Saltwick Bay. The level of the final marker, the double row of nodules forming the Ovatum Band, lies just above the top of the nab but it is visible about 4.5 m above the base of the cliff at the north-west corner of Saltwick Bay: from there it dips south-east in the cliff and appears at shore level in one small area before disappearing beneath beach sand in the middle of the bay.

The Bituminous Shales contain numerous fossils (Fig. 26). Belemnites (*Acrocoelites*) are common and uncrushed, but other fossils are normally flattened and often pyritised. The bivalve *Pseudomytiloides dubius* is common, while dactylioceratid and harpoceratid ammonites include the subzonal form *Harpoceras falcifer*.

In the centre of the bay, near the beach and the path up the cliff, there is a pre-glacial channel with a near-vertical west bank and a gentler eastern slope, entirely filled with boulder clay. It persists seawards as a deep channel (Saltwick Hole) which is particularly noticeable at low tide. On the adjacent shore, many of the sandstone blocks represent the remains of a small quay, built to serve the adjacent alum quarries. There is a large overgrown quarry in Alum Shale on the cliff immediately south of Saltwick Nab and a smaller one south of Black Nab. Heaps of red, burnt shale, overgrown soaking-pits and the ruins of the quay, as well as the size of the quarries, are an indication of the importance of the alum industry here.

Return to Whitby up the cliff path and along the cliff-top walk to the Abbey car park.

3 Excursions: Itinerary 3

Figure 25. East side of Saltwick Nab, Saltwick Bay, Whitby, showing the main marker bands.

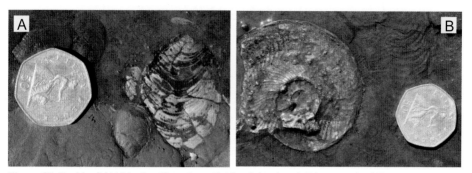

Figure 26. Pyritised (A) bivalve (*Pseudomytiloides dubius*) and (B) ammonite (*Harpoceras exaratum*), bed 43, Bituminous Shales, Saltwick Bay.

ITINERARY 4: ROBIN HOOD'S BAY AND RAVENSCAR

P.F. Rawson and J.K. Wright

OS   1:25 000 Explorer OL 27 North York Moors Eastern area
     1:50 000 Landranger 94 Whitby & Esk Dale
GS   1:50 000 Sheets 35 & 44 Whitby and Scalby

Robin Hood's Bay provides excellent exposures of the Redcar Mudstone Formation and, in the centre of the bay, of Devensian (Pleistocene) tills. The latter yield numerous erratics including boulders of the distinctive Shap Granite (Fig. 27), transported by ice from Shap Fell in Cumbria. At the Peak, and on the shore below, the Peak Fault is seen while to the east of the fault there are good exposures of the Whitby Mudstone Formation. Fig. 28 gives a plan of the geology of the Robin Hood's Bay-Ravenscar area. The Redcar Mudstone and Whitby Mudstone formations should be examined on the rock platform, not at the cliff foot, as the crumbling shale cliffs can fall at any time. There is the very real prospect of being cut off by an incoming tide as eventually it reaches the base of the cliff right around the bay and also to the east of the Peak. The northern exposures in Robin Hood's Bay, from Baytown to Castle Chamber, were described by Senior (1994). More recently,

Figure 27. Erratic boulder of Shap Granite on the shore in Robin Hood's Bay.

Figure 28. (facing page) Map of Robin Hood's Bay and the Peak (Itinerary 4), showing the geology of the rock platform. The outcrop of each of the zones of the Redcar Mudstone Formation is indicated by the colours shown in Fig. 29. Cliff in darker brown shading. Based on data in Howarth (1962, 2002).

3 Excursions: Itinerary 4

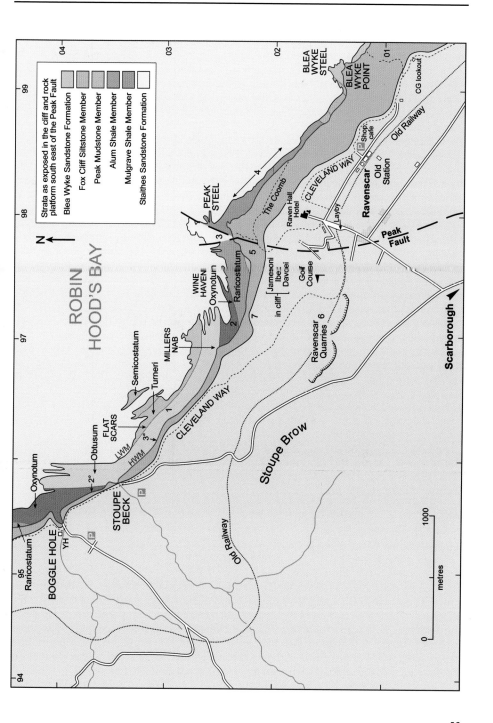

3 Excursions: Itinerary 4

much of this area has been the site of coastal protection works and, due to the subsequent growth of a cover of seaweed and sand, the beach outcrops are now disappointing. This itinerary is thus based on the much better outcrops in the south-eastern part of the bay.

The itinerary follows the succession upwards, starting at Stoupe Beck where there is a small car park and a broad track leading to the beach. Access from Ravenscar is via a single track road with only a few passing places, only suitable for cars or minibuses. It involves a stiff walk back up the Peak, climbing about 150 m, to Ravenscar village and then a walk of about 2.5 km back along the Cleveland Way footpath—or slightly more if Locality 6 is visited. Alternatives are either to return along the shore from the Peak, or to park in the free roadside parking area at Ravenscar, walk down the Peak and take the itinerary in reverse. With more than one vehicle or driver it is possible to shuttle between the two parking areas. The going is reasonably easy on much of the shore, though there is some boulder scrambling in the vicinity of the Peak. Around the foot of the Peak one may encounter a large seal colony.

Robin Hood's Bay is famed for displaying a dome-like structure, the eastern and central parts parts of the dome lying below sea level. The formation of the dome may be a response to Zechstein salt swelling at depth (see section 2). The visible part has been carved by erosion into a huge half-amphitheatre, with the oldest beds cropping out at low water in the centre of the bay. Here they parallel the cliff face, so that the same few beds are exposed for a considerable distance. The only Lower Liassic stage not exposed is the Hettangian, some 90 m of which was met with in nearby boreholes. The cliffs and rock platform expose much of the Sinemurian and the whole of the Lower Pliensbachian sequence, comprising some 150 m of silty and shaly mudstones, siltstones and fine sandstones representing the upper part of the Redcar Mudstone Formation. The Redcar Mudstone succession here has been the subject of considerable study (Van Buchem & McCave, 1989; Van Buchem *et al.*, 1992, 1994; Melnyk & McCave, 1992; Van Buchem & Knox, 1998). Very detailed bed-by-bed logs of the complete sequence were published by Hesselbo and Jenkyns (1996) and Howarth (2002). The latter brought together the late Leslie Bairstow's lifetime work on the sequence and in general the two sets of lithic logs can be matched reasonably.

Buckman (1915) recognised four informal divisions of the Redcar Mudstone Formation in Robin Hood's Bay, and they were used informally in most subsequent publications. However, Howarth (2002, p. 111) formally defined them as members of the Redcar Mudstone Formation:

Ironstone Shale Member	62.73 m
Pyritous Shale Member	26.18 m
Siliceous Shale Member	38.74 m
Calcareous Shale Member	23.35 m visible

The locality map (Fig. 28) shows the distribution of the ammonite zones, as these provide a finer division than Buckman's lithological units. The zones follow Howarth (2002, fig. 4). It is not feasible to reproduce the very detailed bed-by-bed logs published by Hesselbo and Jenkyns (1996) or Howarth (2002), so the interested specialist should refer to those papers for further information. However, we are indebted to Professor Hesselbo for providing a simplified version of his scheme (Fig. 29) which gives a good impression of the whole sequence, including the correspondence between zones and lithology.

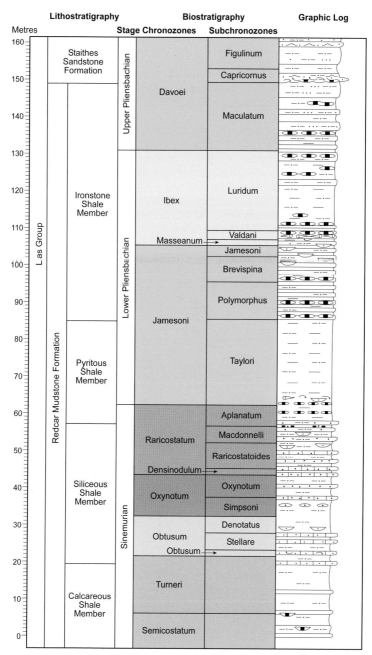

Figure 29. Simplified lithic log of the Redcar Mudstone Formation, Robin Hood's Bay. Kindly provided by Professor S. Hesselbo (Camborne School of Mines, University of Exeter).

3 Excursions: Itinerary 4

Locality 1. Stoupe Beck to Millers Nab (Calcareous Shale and Siliceous Shale members)

It is in the vicinity of Stoupe Beck that the strata parallel the cliff face, so that the same beds are exposed for a considerable distance. On reaching the shore here, proceed south-eastwards over the sandy beach to the low reefs of the Calcareous Shale Member, exposed in the centre of the bay on Flat Scars, and thus the lowest strata seen. This consists of dark grey mudstones interbedded with hard, calcified, silty mudstones. Calcareous concretions are frequent and occasional thin, but laterally quite extensive, shell beds (mainly *Gryphaea*) occur. The characteristic ammonites *Arnioceras* and *Caenisites*, with chambers filled with calcareous mudstone, can be found by careful search of the seaweed-covered reefs. Trace fossils include the vertical U-shaped burrow *Diplocraterion* (Fig. 30).

In the upper rock platform, the prominent reefs of the overlying Siliceous Shale Member can be followed for almost a kilometre south-eastwards until they project into the sea east of Millers Nab. The member gets its name from the repeated occurrence of tough beds of calcareous siltstone and fine sandstone which form the more prominent, higher parts of individual scars. Hesselbo and Jenkyns (1996) placed the base of the Siliceous Shale at the base of their bed 23, which is the first significant sandstone in the succession, forming a prominent ledge opposite Stoupe Beck. This lies a little below the top of the Turneri Zone. The whole sequence was deposited on a shallow shelf, below fair weather wave base. The coarser-grained beds probably accumulated under storm conditions

Figure 30. *Diplocraterion*, a vertical burrow showing spreite in a loose concretion from the Calcareous Shales Member, Robin Hood's Bay.

(see p. 7), and are riddled with trace fossils such as *Rhizocorallium*, *Thalassinoides* and *Teichichnus*, well illustrated by Powell (2010, p. 47). Ammonites are quite common in the more argillaceous, quieter water intervening beds, particularly the zonal form *Asteroceras obtusum*. Large body-chambers of this species can be found weathering out of the shales in the rock platform [NZ 963 032]. Slightly higher, *Gagaticeras* is a common indicator of the overlying Oxynotum Zone; *Oxynoticeras* itself is rare in Yorkshire, but occasional specimens can be found in the scars near Millers Nab.

Locality 2. Wine Haven (Siliceous Shale to Ironstone Shale members)

Rounding Millers Nab, with its waterfall, the small bay at the south-east end of Robin Hood's Bay, known as Wine Haven, is reached. The last thin sandstone marks the top of the Siliceous Shale and, exposed in the rock platform here, is the overlying Pyritous Shale Member. This marks a return to deeper water conditions, consisting of dark grey shales with many calcareous and/or sideritic nodules and with frequent irregular masses of pyrite, small pyritised ammonites and thin-shelled bivalves which lived in dysaerobic (low oxygen) conditions. The best locality for fossils is the rock platform on the corner of the nab as one enters Wine Haven [NZ 972 025], where *Echioceras* (Raricostatum Zone) is common.

In the beach just east of here there is ample evidence of the dock, pier, etc built in Victorian times in connection with the alum works on the cliff top to be visited later. A large channel was excavated in the rock platform (Fig. 31) to enable a boat to be docked here. Square and round, 30 cm diameter holes have been drilled into the rock platform where a pier was constructed and parallel ruts worn on the rock platform show where carts unloaded coal for the alum works and offloaded processed alum. Originally there was a broad track cut obliquely up the cliff, though coastal erosion means that this is barely visible any more.

The section in the Pyritous Shale Member at the base of the cliff in the centre of Wine Haven has been chosen by the International Subcommission on Jurassic Stratigraphy as the Global Stratotype Section and Point (GSSP) for the base of the Pliensbachian Stage. (Meister *et al.*, 2006). The base of the Pliensbachian lies 4.5 m above the top of the Siliceous Shale Member, just above a distinctive row of orange-coloured nodules seen near the base of the cliff. Ammonites are quite common throughout this sequence, though consisting of poorly preserved internal moulds, pyritic phragmocones and crushed body-chambers. The change in fauna which marks the junction is from a fauna dominated by *Echioceras* and *Eoderoceras* to one dominated by *Apoderoceras* and *Phricodoceras*. Please do not attempt to collect specimens from this important section.

In the cliff above, frequent bands of red-weathering sideritic mudstone nodules give the Ironstone Shale Member its name. Upwards of 70 bands of sideritic nodules, some sporadically developed, some continuous bands of siderite, were logged by L.F. Bairstow (Howarth, 2002). Regular changes in climate due to the operation of Milankovic cycles (cycles of variation in Earth's orbit) are thought to be responsible for the regular, pale/dark banding in the lower part of this member, the so-called 'Banded Shales' of Van Buchem *et al.* (1994) and Van Buchem and Knox (1998). Following Powell (2010), the term 'Banded Shales' is used here as an informal subdivision of the Ironstone Shale Member. Six of the ten ammonite subzones of the Lower Pliensbachian are represented in these beds, but the ammonites are poorly preserved and sporadic. Among the genera represented, and which may be found in fallen blocks, are *Platypleuroceras*, *Tropidoceras*, *Acanthopleuroceras* and *Androgynoceras*. The large, semi-infaunal bivalve *Pinna* is common

Figure 31. Dock excavated in rock platform, Wine Haven, Robin Hood's Bay.

at some levels, flattened on the bedding planes, and current-swept belemnite accumulations occur. The Ironstone Shale becomes increasingly silty in the cliff above and grades almost imperceptably into the silty sandstone of the Staithes Formation.

Locality 3. Peak Steel (Peak Fault and Staithes Sandstone Formation)

Proceed over boulders at the south-eastern end of Robin Hood's Bay to reach the Peak Fault (Figs 32, 33), which is 25 m north of the ruins of a pumping station. Redcar Mudstone is faulted here against Whitby Mudstone. A zone of crushed shale 10–20 cm wide in the base of the cliff marks the fault plane. The Bituminous Shales, the upper, informal subdivision of the Mulgrave Shale Member, are faulted against the Ironstone Shale of Jamesoni Zone age containing bands of reddish-weathering ironstone concretions. This indicates a throw of 153 m.

Where the Peak Fault bifurcates on the shore, the two branches bound a prominent triangular outcrop of Staithes Formation known as Peak Steel. The 'steel', which extends seawards for nearly 500 m at low water and is covered by high tide, is stepped between Redcar Mudstone Formation to the west and Whitby Mudstone Formation to the east. The reefs are mostly covered with barnacles, but the extensive sandstone surface shows hummocks and swales characteristic of hummocky cross-stratification. The sequence is best examined nearest the cliff where cover is minimal. Here, sideritic sandy limestone with profuse bivalves (*Protocardium truncatum*) alternates with the fine-grained, laminated, cross-bedded sandstone. To the east of the fault the Jet Rock (lower part of the Mulgrave

Figure 32. The Peak Fault from the air. (Photo kindly provided by William Deadman.)

Shale Member) is exposed among the boulders, the Top Jet Dogger level being represented by a calcareous mudstone very different from its development west of the fault.

The Bituminous Shales are well exposed south-east of the fault. They are laminated and silty, and the shells of ammonites (*Harpoceras falcifer*, *Dactylioceras* spp.) are completely flattened. Though scattered throughout, the bivalve *Pseudomytiloides dubius* is abundant in layers a few centimetres apart.

Locality 4. The shore south of Peak Steel (Mulgrave Shale and Alum Shale members)

Continue southwards for 250 m across the Bituminous Shales to reach a small bay where the Peak Stones form one of the few marker beds in the Bituminous Shales (Fig. 34). These are discoidal concretions up to about 1.5 m diameter that cap low stacks and are known locally as 'fairies' tables'. They show cone-in-cone structure, a post-depositional effect. Although these are similar to the concretions in the Top Jet Dogger to the west of the fault, they are here about 7.5 m above the equivalent level.

The upward succession is almost continuously exposed for some distance in the cliff and in the mid-shore. Between 400 and 500 m south of the Peak Fault, the Ovatum Band (top of the Mulgrave Shale Member), a double bed of pyrite-skinned concretions associated with calcareous lenses exhibiting cone-in-cone structure, is clearly visible in the cliff. It descends gently down to the base of the cliff 500 m south of the Peak Fault and here it is well exposed in the rock platform. This is one of the few marker beds in the Mulgrave Shale Member that occurs both sides of the Peak Fault. As well as occasional *Ovaticeras pseudovatum*, *Dactylioceras* spp. and *Phylloceras* sp. are also present in the concretions.

3 Excursions: Itinerary 4

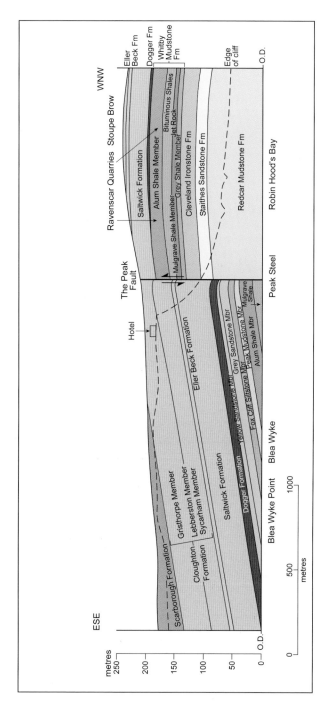

Figure 33. Diagrammatic section across the Peak Fault at Ravenscar, as seen from the shore.

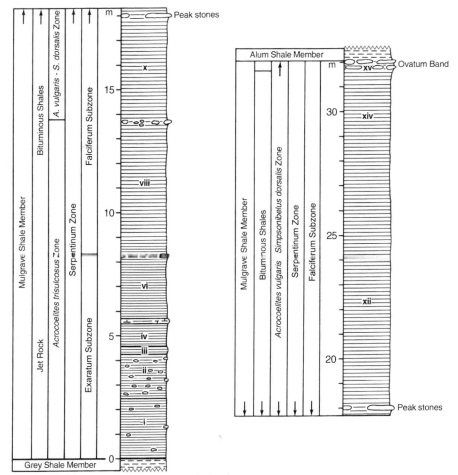

Figure 34. Lithic log of the Mulgrave Shale Member at Peak. For lithological key see Fig. 15. Modified from Rawson and Wright (1996, fig. 18), based mainly on data in Howarth (1962).

The succeeding Hard Shale unit of the Alum Shale Member is well exposed for a thickness of 4 m above the Ovatum Band. Like the Bituminous Shales, it contains flattened ammonites and *Pseudomytiloides dubius*, but it is not laminated. *Hildoceras bifrons* replaces the *Harpoceras falcifer* of the Bituminous Shales, and *Dactylioceras* spp. and belemnites are common. Beyond this exposure there is a boulder-strewn area and a low grassy undercliff and those unwilling to tackle this stretch are advised to return to the Peak, where a steep path leads up the cliff, or if preferred and the tide allows return along the shore to Stoupe Beck.

Proceeding southwards, after some 200 m of boulder-strewn beach, the rock platform and cliff for the next 150 m expose almost the whole 37 m thickness of the Alum Shale Member. The first 8.7 m belong to the Commune Subzone. *Dactylioceras commune* is common, with *Dactylioceras* spp. and *Hildoceras sublevisoni*. The succeeding 5 m belong to the Fibulatum Subzone. *Peronoceras fibulatum* and *Peronoceras* spp. are common, along with *Hildoceras bifrons*. The bivalve fauna has changed from that seen previously in the Bituminous Shales. Infaunal, bottom living forms such as *Dacromya ovum* and *Gresslya donaciformis*, in life position, are commonly found. Bottom living conditions during deposition of the Alum Shales clearly were not anoxic.

This lower part of the Alum Shale Member shows only a limited development of calcareous concretions, but at the southern end of the exposure the succeeding 12 m of shales contain frequent bands of concretions. These form the lower part of the cliff in the southern half of the exposure, and descend into the rock platform. This more calcareous facies of the Alum Shale Member is known informally as the Cement Shales. It belongs to the Crassum Zone, and *Catacoeloceras crassum* is abundant, along with *Hildoceras bifrons* and *Pseudolioceras lythense*. The sea floor was becoming more oxygenated with time, and *Gresslya donaciformis* and *Dacromya ovum* are joined towards the top of the Cement Shales by small, well-preserved *Trigonia literata*, again all in life position.

The exposures further along the shore from here are scattered between extensive boulder-strewn areas and access is more difficult and can be dangerous. Thus, the recommendation is not to proceed further, but at the northern end of the Alum Shale outcrop [NZ 985 019] pick up a path which climbs the low undercliff and then winds its way through The Combe to join the main path back to Ravenscar.

For those who wish to examine these higher beds, it is best to park at Ravenscar and walk down to the shore at the Peak then work south-eastward towards Blea Wyke. Beyond the exposures described here there is a gap in exposure before the remainder of the Alum Shale Member is visible (see Fig. 35, and Howarth, 1962). The overlying Peak Mudstone and Fox Cliff Siltstone members and the highest Lower Jurassic sequence, the Blea Wyke Sandstone Formation, were described by Dean (1954) and Knox (1984).

Locality 5. The Peak (Staithes Sandstone Formation and Peak Fault)

Near the top of the path is a viewpoint overlooking Robin Hood's Bay. Pause by the fence at the cliff edge for the classic view of the Robin Hood's Bay Dome, with the reefs of Redcar Mudstone swinging round through 180° (Fig. 36). The crest of the structural dome lies seaward of the coastline within the bay, but the truncated beds on its flanks form curving scars for a distance of five kilometres to the north-west. Below, small faults branch out from the main Peak Fault, which splits into two smaller faults across the shore.

Just beyond the National Trust sign, the Staithes Sandstone Formation (Upper Pliensbachian) is seen on the right-hand side of the path. Micaceous sandstones, siltstones and a thin, nodular ironstone are present and yield Margaritatus Zone fossils. On the left-hand side of the path is a gully marking the line of the Peak Fault (Figs 32, 33). On the eastern side of the fault there is a continuous sequence of strata beginning with an attenuated exposure of the Yellow Sandstone Member (Upper Toarcian) succeeded by the heavily jointed Dogger Formation with pebble beds, and by massive sandstones, siltstones and a thin coal of the Saltwick Formation. Looking back at the Middle Jurassic beds note that although they are on the downthrown side, the Dogger being downthrown 90 m relative to its position on the west side of the fault, the beds dip down into the fault. This suggests that the fault plane curves at depth (i.e. that it is listric). The fact that the throw in the Liassic beds at the bottom of the cliff is 63 m greater than the throw of the Dogger at the top of the cliff is a clear indication of the amount of erosion which affected the western side of the fault prior to or contemporaneous with the deposition of the Dogger. Much of the Upper Toarcian sequence is absent west of the fault due to this erosion (Loc. 6).

3 Excursions: Itinerary 4

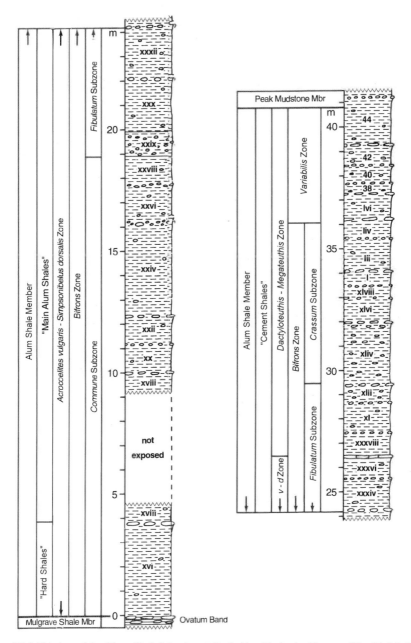

Figure 35. Lithic log of the Alum Shale Member at Peak. For ithological key see Fig. 15. Modified from Rawson and Wright (1996, fig. 19) based mainly on data in Howarth (1962).

3 Excursions: Itinerary 4

Figure 36. Robin Hood's Bay from the Peak. The curvature of the scars reflects the Robin Hood's Bay Dome.

The Peak Fault is now known to mark the western margin of the Peak Trough (Milsom & Rawson, 1989). To the east, within the trough, a more complete Toarcian sequence is preserved than is visible anywhere else along the coast. Movement on the fault probably started during the Toarcian and according to Alexander (1986), the fault was active, downthrowing to the south-east, during Mid Jurassic times, allowing a thick sequence of fluvial sands to accumulate in the subsiding area close to the fault.

Proceed up the path which climbs up across the golf course towards the Raven Hall Hotel and then to the main road. From the entrance to the hotel proceed down the track heading south-westwards (Cleveland Way) past the National Trust visitor centre. From here one can continue along the Cleveland Way until it meets the road down to Stoupe Beck. Alternatively, carry on down the main track and, after about 250 m, fork left at a signpost for 'Old Brick Works' onto the former railway line, now a footpath and cycle route. About 400 m along is a National Trust sign 'Brickyard. Alum Quarry' [NZ 975 015] and an adjacent gate provides access to the site.

Locality 6. Ravenscar brickworks (Whitby Mudstone, Dogger and Saltwick formations)

The quarry was initially set up in the 17th Century to quarry Alum Shale for the nearby alum works, but long after the alum works ceased production in 1860, a brickworks was sited in the quarry in 1900. Part of the site has been cleared to provide access to the brick kilns: beyond these it is possible to walk between trees and undergrowth for about 20 m to reach an area immediately in front of the quarry face that is quite clear of vegetation.

The succession here lies 450 m west of the Peak Fault and provides a marked contrast with the succession on the east side, between Peak Steel and Blea Wyke Point. The high, inaccessible part of the quarry face shows a thin (1 m maximum) development of the Dogger Formation resting directly on the Main Alum Shales (Alum Shale Member). Some 57 to 58 m of beds, including the Cement Shales, the Peak Mudstone and Fox Cliff Siltstone Members and the Blea Wyke Sandstone Formation, are absent on this western, upthrown side of the fault.

The Dogger is overlain by level-bedded alternations of shale and sandstone of the Saltwick Formation. Many blocks of the Dogger have fallen to the quarry floor. Some contain bored pebbles of derived Liassic nodules. *Diplocraterion* (U-shaped burrows), some superbly developed, penetrate the full thickness of the Dogger Formation from its top surface, being filled with coarse quartz sandstone. Many blocks of plant-rootlet sandstone have fallen from the Saltwick Formation and from here Sargeant (1970, pl. 21) figured a dinosaur footprint, referred to *Satapliosaurus*. South-east of the quarry, close to the Peak Fault, the Dogger Formation thins out entirely. The younger beds also show a marked reduction in thickness from east to west across the fault, the Saltwick Formation being reduced from at least 57 m to only 30 m and the Cloughton Formation from 77 m to about 55 m. The differences in the successions on the two sides of the Peak Fault are shown in the cross section (Fig. 33).

It is now generally accepted that the omission of nearly 60 m of Lower Jurassic beds, and the attenuation of the Ravenscar Group by some 50 m, is due to contemporary Jurassic movement on the Peak Fault allowing Upper Toarcian beds to be preserved on the downthrown side, while on the shallow upthrown side they were either scoured off or not deposited. Milsom and Rawson (1989) traced the Peak Fault Zone into the offshore area and defined a Peak Trough, a zone of trough-faulting 5 km wide bounded to the west by the Peak Fault and to the east by a series of faults running from Scarborough to Red Cliff. Normal faulting took place in the trough in Triassic, Mid Jurassic, probably Late Jurassic/Cretaceous and Cenozoic times. Movement could have been by gentle creep rather than sudden earthquake shock for much of the time. The surface expression of the trough may have been only a metre or two (Alexander, 1986) and as such the trough operated at Ravenscar for much of the Aalenian and Bajocian, allowing a succession to accumulate within the trough 60 m thicker than that on the western flank. Powell (2010, fig. 27) shows the northern extension of the faulting into the North Sea, showing much thicker accumulations of Jurassic strata within the trough.

The Peak Fault Zone now takes its place amongst the many other extensional Mesozoic synsedimentary fault systems which have been discovered in the North Sea and in southern England. Strike-slip faulting on the fault system only took place much later, during the Alpine compressive movements also responsible for the formation of the Cleveland Anticline (Milsom & Rawson, 1989). A structural analysis of the Peak Fault Zone as it extends south to Cayton Bay shows that it could be interpreted as having operated during these compressive movements as a transcurrent fault system of dextral movement. The main north-westerly trending faults and the subsidiary NNE/SSW trending branch faults fit naturally a near north-south primary stress pattern. ESE/WNW trending secondary drag folds fitting this stress pattern can be recognised at Osgodby Point (Wright, 1968, fig. 9) and at Hayburn Wyke.

3 Excursions: Itinerary 4

Locality 7. Peak Alum Works

Return to the Cleveland Way and walk northwards to reach the site of the Peak Alum Works. Alum was made here between 1650 and 1860, using Alum Shale brought down from the quarry at Locality 6. This well-preserved site is excellently managed by the National Trust, with numerous boards explaining the processes involved in extracting the alum.

The route back to the Stoupe Beck car park now follows the Cleveland Way for 2 km, with spectacular views of Robin Hood's Bay.

3 Excursions: Itinerary 5

ITINERARY 5: CLOUGHTON WYKE TO THE HUNDALES

J.K. Wright

OS   1:25 000 Explorer OL 27 North York Moors Eastern area
        1:50 000 Landranger 101 Scarborough
GS   1:50 000 Sheet 35 & 44 Whitby and Scalby

This itinerary provides a traverse through part of the Ravenscar Group from the Lebberston Member (Millepore Bed) to the Moor Grit (Scalby Formation), showing in detail two of the marine incursions into the Ravenscar Group, the Cloughton and Scarborough formations, and the intervening and succeeding marginal-marine or non-marine sequences of the Gristhorpe Member and the basal Scalby Formation (Figs 37, 38). The full excursion involves a 2 km walk along the rock platform at the base of the cliffs and over stretches of beach boulders. The going can be quite strenuous. It is possible to drive cars or minibuses down to the car park at the Salt Pans on the northern side of Cloughton Wyke [TA 019 952] (Fig. 37), though the car park is small (half a dozen vehicles) and the road narrow, and large parties will have to be dropped off in Cloughton.

Locality 1. Cloughton Wyke (Lebberston Member to basal Scarborough Formation)

1A. The north side of Cloughton Wyke. Descend to the shore from the car park which is situated 500 m north of the centre of Cloughton Wyke and walk northwards over large fallen sandstone blocks (some of which show excellent sedimentary structures) for 250 m. Here the Lebberston Member, represented by the 3 m thick Millepore Bed, forms a series of reefs running south-eastwards out to sea. The bed consists of three tiers of shelly, calcareous sandstone and sandy limestone each forming a coarsening-upwards cycle. The middle tier is the most fossiliferous and the upper part is extremely hard where is is cemented by siderite. The fauna includes *Arcomya elongata, Lima duplicata, Entolium demissum, Pleuromya beani, Trigonia* sp., *Pholadomya saemanni* and *Modiolus imbricatus*. The branching bryozoan *Haploecia [Millepora] straminea*, which originally gave its name to this unit, is less common than at Osgodby Point (Itinerary 9). Many of the shells are preserved in the white-weathering clay mineral dickite. Excellent cross-bedding makes the upper tier very distinctive. Compared with the sections south of Scarborough (Itinerary 7), the Millepore Bed here shows a marked decrease in the content of ooids and shell debris and an increase in calcareous sandstone and particularly iron carbonate, indicating that it accumulated nearer the shoreline (Bate, 1959).

The succeeding informal stratigraphic unit named the Yons Nab Beds by Bate (1959) can be regarded as the basal, quasi-marine part of the Gristhorpe Member. The Yons Nab Beds are exposed in the low cliff on the way back towards Cloughton Wyke. Here quasi-marine beds interdigitate with fully non-marine beds. Resting on the Millepore Bed is 0.6 m of ripple-drift cross-laminated sandstone, succeeded by 1.6 m of flaser-bedded, intertidal, shaly sandstone with numerous bivalves including *Trigonia, Pecten* and *Ostrea*. A variably bioturbated, ripple-drift cross-laminated sandstone (0.6 m) seems to complete the Yons Nab Beds sequence. It is succeeded by 1 m of shale with coaly, carbonaceous layers. This unit weathers back readily.

However, traversing southwards across a large rockfall, a low cliff is reached exposing, above the carbonaceous shale, a further 2.6 m of quasi-marine beds, an almost identical situation to that seen at Yons Nab (Itinerary 9). These consist of flaggy and well-bedded

3 Excursions: Itinerary 5

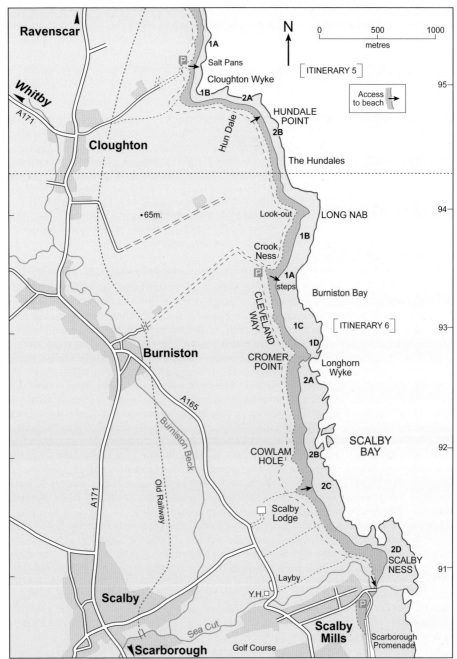

Figure 37. Map of localities between Cloughton and Scalby (Itineraries 5 and 6). Cliff in darker brown shading.

3 Excursions: Itinerary 5

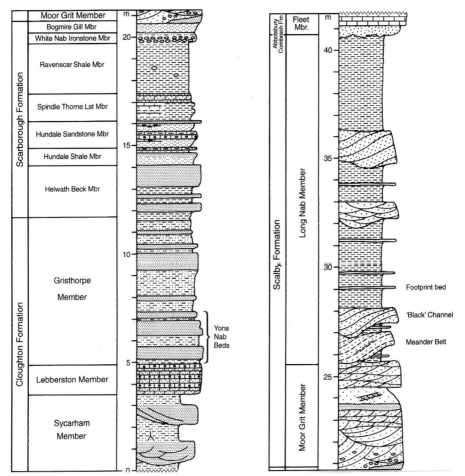

Figure 38. Lithic log of the Ravenscar Group between Cloughton Wyke and Scalby Ness. For lithological key see Fig. 15. Modified from Rawson and Wright (1996, fig. 9).

sandstone containing occasional *Diplocraterion*, and are strongly bioturbated towards the base. These beds were first included within the Yons Nab sequence by Livera and Leeder (1981). The top of this unit forms a slippery ledge facing southwards and above it occurs a fully non-marine succession of 4.5 m of shales with thin coals and a rootlet bed, overlain by thick sandstone.

1B. Cloughton Wyke. The non-marine sandstones and shales of the upper part of the Gristhorpe Member are exposed in the cliffs and rock platform for the 0.5 km from the car park area southwards to the centre of Cloughton Wyke. The lower beds, occurring just above the sequence described at Locality 1A, are exposed in the cliffs on either side of the boulder-strewn beach below the car park. They consist of thick, planar-bedded sandstones, a facies unusual in the non-marine beds and which is almost certainly the result of sheet flooding. This can be demonstrated by a close examination of these beds in the cliff between the car park and Cloughton Wyke. **The cliff is, however, unstable with considerable overhangs** (Figs 39, 40), **and great care should be taken along this section**.

The planar-bedded sandstones generally occur in four approximately half-metre tiers capped by a thin coal (Fig. 39). Below the lowest tier there is a gradual shallowing-upwards sequence extending over a thickness of 1.5 m from shale with siderite nodule bands (the Cloughton Wyke Plant Bed) through striped silty shales into a thin-bedded, ripple-marked sandstone. The basal tier of massive sandstone infills well-defined grooves and flutes (gutters) carved into the thin-bedded sandstone. The bottom structures indicate a current flow from NNE to SSW. The thick beds of sandstone thus appear to be associated with a large distributary channel to the north-east, which was subject to repeated breaches such that extensive sheet sands (crevasse splays) spread south-westwards into the Cloughton Wyke area. Within each tier, though the bedding is frequently planar reflecting the strong current flow, ripples and cut-and-fill structures are also seen. Excellent sedimentary structures are to be found in the fallen blocks beneath the cliffs. After each episode of crevasse splaying, the low-lying area around Cloughton Wyke was then colonised by marsh plants. Well-preserved rootlets are visible in tiers 2 and 4, the latter lying immediately beneath the thin coal, and another rootlet horizon lies immediately above the coal (Fig. 39).

The coal forms an excellent marker horizon and can be followed, dipping gently southwards, into the centre of Cloughton Wyke. Here the top of the underlying fourth tier, heavily rooted, forms the upper rock platform. Above the coal come 6 m of alternations of sandstone and shale. The sandstone resting on the coal is cross-bedded, though laterally persistent, and it is heavily lode casted or even convoluted into the underlying coal and carbonaceous shale. Numerous 2 cm diameter sand-filled tubes extend down into the coal. These presumably are root structures.

There follows 2 m of alternations of shale and sandstone, often heavily rooted and convoluted. Above is a 3.5 m thick pale grey, alluvial clay which immediately passes laterally into a channel sandstone, deposited rapidly and convoluted at its base. The top of the clay is rooted and overlying it is a dark, laminated, shaly siltstone. The boundary between the two beds is clearly erosive on a small scale and infilled burrows carry dark silt several centimetres down into the underlying grey clays. Gowland and Riding (1991) took this surface to mark the base of the Helwath Beck Member of the Scarborough Formation, and it thus marks the beginning of a marine transgression.

The shaly siltstones at the base of the Helwath Beck Member pass up into massive, convoluted sandstone; the combined thickness is about 2 m. The convolutions are thought to have been caused by an earthquake shock due to movement on the Peak Fault, here running 1.6 km to the west. The bed is remarkably persistent and can be traced all the way to Ravenscar. The upper part of the member, comprising 4 m of extremely massive sandstone containing excellent marine trace fossils, forms the middle part of the cliff in the centre of the wyke (Fig. 40).

Amid the boulders on the shore, the medium-grey, micaceous, silty clays of the Cloughton Wyke Plant Bed can be traced from the base of the cliffs south of the car park to the centre of Cloughton Wyke, where they go below sea level. The plant bed is seen at its best in the middle of the rock platform half way between these localities. The flora is rich, usually carbonised, but occasionally preserved in a flexible, chitinous form. Individual species have a remarkably local distribution. They include *Ptylophyllum pecten*, *Cladophebis* sp., *Czechanowskia* sp., *Nilssoniopteris* sp. and *Otozamites* sp.

Figure 39. (A) Cliff section in the Gristhorpe Member on the south side of Salt Pans, Cloughton Wyke, immediately below the car park. GH = gutter horizon; RB = rootlet bed. (B) Rootlets in the middle rootlet bed.

3 Excursions: Itinerary 5

Figure 40. The Scarborough Formation, cliffs east side of Cloughton Wyke. The sandstones and shales of the Helwath Beck Member form the lowest third of the cliffs. They are overlain by the thin-bedded alternations of impure limestone and shale of the Hundale Shale, Hundale Sandstone and Spindle Thorne Limestone members. These are capped here by grass-covered Late Devensian tills.

Locality 2. Hundale Point (Scarborough and Scalby formations)

2A. North side of Hundale Point. About 70 m along towards Hundale Point, the junction of the Helwath Beck Member resting on the Gristhorpe Member is seen in the base of the cliff. The undulating, erosive contact of dark, silty shale on grey clay is well seen, dark, silt-filled burrows extending down into the Gristhorpe Member. In the flaser laminated, silty sands 0.5 to 1 m above the contact *Diplocraterion* occurs. Just beneath the convoluted bed, hummocky cross-bedding and climbing ripples in laminated sandstone indicate tidal scour in intertidal conditions.

After a further 30 m of scrambling over beach boulders, a wave-cut platform in the channel sandstone referred to in Locality 1B is reached. The going is easier now. The massive upper sandstone of the Helwath Beck Member forms the lower middle of the cliff and huge blocks of this sandstone litter the foreshore. The beautiful, laminated, scour-and-fill cross-bedding is well seen in weathered blocks, with numerous *Diplocraterion*, especially in the top 0.3 m.

Yet more scrambling over beach boulders is necessary to reach the top of the Helwath Beck Member, which reaches the rock platform just east of Hundale and strikes out to sea, forming a small reef named Hundale Scar. Standing on the top of the Helwath Beck

Member, with its profusion of U-shaped burrows, one can see the stratigraphy of the remainder of the Scarborough Formation displayed in the rock platform and cliff as set out by Gowland and Riding (1991).

The Hundale Shale Member (2.6 m) crops out between the boulders up to the base of the cliff. Its name is something of a misnomer, as it comprises intensely bioturbated silty sandstone in the lower part, passing up into argillaceous sandy limestone with *Gervillella*. The top is marked by a thin, distinctive, red-weathering, iron-rich bed.

The Hundale Sandstone Member (4.0 m), seen in the lower cliff, consists of two contrasting tiers of sandstone. The lower one is thin bedded, flaggy and argillaceous, and intensely bioturbated, with *Rhizocorallium* and *Diplocraterion*. It is overlain by a more massive tier of sandstone which is less argillaceous. A thin bed of pink-weathering, sideritic limestone resting on shale separates the tiers. The top surface of the Hundale Sandstone is very shelly, with *Lopha* and *Pentacrinites* (compare with the Crinoid Grit, the lateral equivalent of these beds in the west of the Cleveland Basin).

The Spindle Thorne Limestone (3.7 m) consists of regular alternations of superbly bioturbated sandy shale with bioclastic, argillaceous limestone containing large bivalves. It descends gently to form the base of the cliff for several tens of metres round to Hundale Point.

The Ravenscar Shale Member (8.2 m) consists of a thick sequence of dark-grey shales with small, fossiliferous concretions.

The White Nab Ironstone Member (1.3 m) is thinly developed here. It consists of sulphurous, grey, sandy shales with three layers of iron-rich concretions.

The Bogmire Gill Member (approx. 2.5 m) is much more marginally marine, consisting of siltstone passing up into fine-grained sandstone. It is abruptly overlain by the cross-bedded sandstones of the Moor Grit.

The upper three members of the Scarborough Formation can be examined near the base of the cliff at Locality 2B.

2B. The cliffs south of Hundale Point. Immediately south of the path which leads from Hundale Point up to the cliff-top path, the Ravenscar Shale forms the base of the cliff for several tens of metres. The small, fossiliferous concretions in the grey shales yield frequent ammonites, including *Teloceras* sp. and *Dorsetensia* sp. The White Nab Ironstone can be a little difficult of access, tending to be hidden by scree. The inter-tidal, flaser-bedded silts and fine-grained sandstone of the Bogmire Gill Member can be reached by climbing the grassy slopes beneath the Moor Grit cliffs.

The Scarborough Formation passes rapidly beneath beach level southwards, and the rock platform and cliffs all the way to Scarborough are composed of Scalby Formation sandstones and shales, overlain by glacial till. At one time, the basal unit of the Scalby Formation, the 6 m thick sandstone unit comprising the Moor Grit Member, formed the lower cliff for several hundred metres south of Hundale Point. However, recent rock falls have obliterated much of this spectacular sequence of 6 m high cross-bedded units. A drawing of this cliff section as it appeared then was given by Nami and Leeder (1978). A 50 m long section in the Moor Grit is still present above Hundale Point however. The coarse, mature sandstones are rich in coalified plant remains with subordinate charcoal fragments. The massive channel sands alternate with and pass laterally into off-bank laminated siltstones and crevasse-splay sheet sands which alternate with shale. The impression is of a rapidly accumulating series of beds, with little reworking or lateral migration of channels. The Moor Grit was interpreted by Ravenne (2002) as being formed by the nesting of rectilinear fluvial channels with sand fill set in a broad palaeo-valley, excavated during a relative fall in sea level. The Moor Grit has been divided into 'Prisms' I and II by Eschard *et al.* (1991), Ravenne (2002) and Hesselbo *et al.* (2003), Prism I being here at Hundale and Prism II being situated in a separate palaeo-valley south of Scarborough (Itinerary 8).

The route back to the Salt Pans is to return to the Ravenscar Shale outcrop. A sloping, grassy cliff has formed here on the outcrop of this unit, giving an easy ascent up to the Cleveland Way, which follows the cliff top back to the car park.

ITINERARY 6: BURNISTON AND SCALBY BAYS

J.K. Wright

OS   1:25 000 Explorer OL 27 North York Moors Eastern area
         1:50 000 Landranger 101 Scarborough
GS   1:50 000 Sheets 35 & 44 Whitby and Scalby

This itinerary demonstrates the remarkable variety of alluvial facies present in the Scalby Formation. It was the most prolonged of the non-marine episodes during the accumulation of the Ravenscar Group that saw the extensive development of meander-belt sands on a broad alluvial plain, with dinosaur footprint beds and beds with well-preserved plants. Parking is available for cars and minibuses only at Crook Ness car park, above Burniston Bay [TA 026 935] (Fig. 37). This is essentially a low-tide itinerary, as access from Burniston Bay south to Scalby Bay around Cromer Point is only possible within 3 hours of low tide.

Locality 1. Burniston Bay (Scalby Formation)

The footpath and concrete steps lead to the beach. The rock platform throughout this bay and the succeeding Scalby Bay is cut in the upper surface of a meandering river complex known as the meander belt unit of the Long Nab Member (Nami, 1976). The meander belt succeeds the Moor Grit seen in Itinerary 5, and marks the point where the strong currents which had deposited the Moor Grit ameliorated and a broad, shallow valley was infilled across its full width by sediments laid down by a meandering river draining the higher ground to the north. For three kilometres one can walk over the sediments with their excellently displayed fluvial structures. It should be noted that the course of the valley occupied by the meander belt runs obliquely to the present cliff line and that the width of the infilled valley is only 2.5 km. It effectively followed the course of the Peak Trough (Fig. 3) and its course may have been determined by penecontemporaneous movements in these faults (Alexander, 1986).

Extensive work has been done on the meander belt unit in recent years and this can only be summarized here. First discovered by Nami (1976), the principal subsequent works are by Eschard *et al.* (1991), Alexander (1992), Ravenne (2002) and Ielpi and Ghinassi (2014). With increasing understanding, it has become possible to apply many of the terms used by geographers when discussing present-day meander belts to this Jurassic one.

In a meander belt, the stream passes across from one side of the meander plain to the other in a gently curving course, reaches an inflection point and curves sharply back through a meander loop, often through 180 degrees, before continuing its course back across the plain. As the stream erodes the outer bank of the meander loop, and the channel migrates in this direction as well as downstream, sediment is deposited on the inner bank, forming a point bar. The bar is thus made up of beds of sediment, which may dip at 20 degrees down into the channel, and the internal ripple-drift structure of these beds may show the direction of water flow, at right angles to the principal dip direction.

The plan of the migrating stream shows the upper curved edges of the point-bar deposits marking the history of migration of the channel in a series of meander scroll bars. The stream channel is confined by levees, composed of ridges of gently dipping sand and mud built up at high stages of stream flow. Breaks in the levee at very high stream flow can result in water pouring out in a crevasse splay, depositing planar or gently inclined beds of plane-parallel laminated sand.

Over time, with downstream migration of the point bars, the stream will cut into and erode older point-bar deposits, with reactivation surfaces, so that a complex series of intersecting scroll bars can result. This is the situation in Burniston Bay and the interested reader is referred to Ielpi and Ghinassi (2014) for a description of the meander belt here. The meander belt is much more easily studied in Scalby Bay (Loc. 2) and is described below. There are many features of interest in the Long Nab Member above the meander belt and these are best seen in Burniston Bay.

1A. Burniston Steps. As one reaches the low cliffs on either side of Burniston Steps one comes repeatedly to sections showing such similar features that a single sketch (Fig. 41) can be used to illustrate them. Above the meander-belt sandstones in the rock platform come 2 to 3 m of grey, alluvial clay with subordinate silty sandstones. One, and frequently two beds of dark, carbonaceous clay are present full of flattened plant stems. Rootlets pass down into the meander-belt sandstone. Sideritisation in the clay and upper meander-belt sandstone is intense, both as large concretions and as beds of sphaerosiderite. These ironstones are interpreted as evidence for the existence of ancient soil profiles (the palaeosols of Kantorowicz, 1990). Siderite in such quantities would be expected to precipitate from oxygen-deficient groundwater beneath peaty soils, represented here by the carbonaceous clay.

Thin alternations of sheet sandstone and silty clay follow, with the development of at least a dozen localised channels which follow a linear course across the rock platform, frequently eroding the meander-belt sands. These were first described by Black (1928), and are known by Black's letters A to H. In almost all of the low cliff sections which cut across these channels, one can demonstrate that level-bedded alternations of siltstone and sandstone pass laterally into strongly cross-bedded channel sandstones, with the channels cutting through the alluvial clays and frequently resting on meander-belt sandstone.

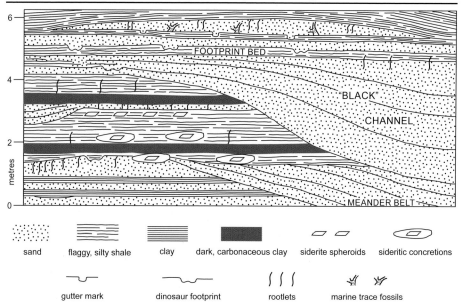

Figure 41. Diagrammatic section of the lower Long Nab Member, Burniston and Scalby bays.

Just above the horizon at which channels develop a ledge of sandstone projects out, on the underside of which moulds of dinosaur footprints are occasionally seen. Dinosaur footprints at this locality were first described by Hargreaves (1914), introducing the term Burniston Footprint Bed. Whyte *et al.* (2010) show the horizon at which the footprints occur excellently in their fig. 9. However, well-preserved footprints are only occasionally visible here, and the best footprints have been found at locality 1B.

1B. North end of Burniston Bay. Proceed over the boulder beach or rock platform towards Long Nab, the type-section of the upper member of the Scalby Formation. The Nab is protected from erosion by several huge blocks of cross-bedded sandstone up to 4 m thick derived from the lower part of the Long Nab Member. The cliff section on the south side of Long Nab in Burniston Bay displays well the main features of the higher part of the member. There are alternations of fine clays and siltstones with occasional sand-filled channels. A small channel is seen in section at eye level, cutting down half a metre into the underlying shale. The sandstones above are more persistent and even bedded. At a height of 4 m above the rock platform occurs the Burniston Footprint Bed. The dinosaur prints were made in soft, silty clay subsequently infilled by a gentle incursion of silt and sand, preserving the prints as casts. As blocks of this silty sandstone fall to the beach they frequently come to rest upside down at the foot of the cliffs, displaying the footprints very well (Fig. 42). Two types were figured by Black *et al.* (1934) and further descriptions of footprint finds in the Scalby Formation were given by Ivens and Watson (1994). A discussion of the various preservational styles of these footprints and the type of animal which

Figure 42. Saurian footprints from a fallen block of the Burniston Footprint Bed, Burniston Bay.

may have made them was published by Romano and Whyte (2003). Whyte *et al.* (2006) figured a single footprint 55 cm long and 40 cm wide which they suggested may have been made by *Megalosaurus*.

1C. The south end of Burniston Bay. Return now to Burniston Steps and work southwards. After a lengthy traverse of fallen blocks of channel sandstone, one comes to an area of regularly bedded sandstone in the low cliff, overlain by clay and shale. A 60 to 70 cm thick sandstone bed here has a quasi-marine appearance. Fallen blocks show that it consists of well-sorted, ripple-drift laminated, fine-grained sandstone. The top surface is riddled with 1 cm diameter burrows now filled with clay from the overlying bed. These easily weather out, leaving networks of burrows attributed to the marine trace fossil *Ophiomorpha* by Livera and Leeder (1981). Where blocks have turned over in falling from the cliff, the undersides of the blocks reveal infilled scour channels or gutter marks. The slightly sinuous channels are 20 cm deep and 25 cm wide. The upturned blocks on the beach show that the scoured channels were rapidly infilled by cross-bedded sand and that then ripple-drift bedded sand built up, eventually to be colonized by the organism responsible for the *Ophiomorpha*. The quasi-marine bed occupies no more than 120 m of the cliff section at this end of Burniston Bay and appears to be a lens almost certainly infilling a shallow channel scoured by a tidal surge. Thin, localised deposits of a similar nature occur elsewhere in Burniston and Scalby bays at the same horizon.

1D. Cromer Point. At Cromer Point, Black's Channel E is seen well, there being a rapid lateral transition from overbank laminated siltstones and sandstones into massive channel sandstone. Current action was so strong that rip-up clasts of laminated sandstone several tens of centimetres across occur here (Fig. 43).

Having rounded Cromer Point it is necessary to negotiate Longhorne Wyke. The cliff on the south side of this gully falls sheer into the sea for much of the tidal cycle and it is not possible to get round it within 3 hours either side of high tide. If the tide is in, a return to Burniston Steps may be necessary.

Locality 2. Scalby Bay (Scalby Formation)

South of Cromer Point, within the long stretch of Scalby Bay, there is a succession of small bays cut in the soft siltstones and clays of the Long Nab Member, with promontories in between formed by Black's sand-filled channels projecting from the cliff. Cliff falls take place regularly here and care should be taken.

2A. The beach and cliff south of Longhorne Wyke. As one proceeds south from Cromer Point cross-bedded meander-belt sandstones are excellently displayed in the rock platform. The meander scrolls curve gently round from southerly to easterly (Fig. 44). Small-scale cross-bedding within the massive, southerly dipping beds, particularly small-scale trough cross-bedding, shows that the current flow was eastwards, with the meander loop migrating southwards.

On reaching the large pile of sandstone blocks marking the next channel [TA 029 925], two small sandy bays are seen, with channels F and G forming promontories. For 100 m south of the sandstone blocks, the cliff section reveals a whole series of small, sand-filled channels stacked one above the other, some choked with the carbonised remains of tree trunks, and with much slumping. This complex series of beds passes laterally into sheet sandstones and shales with occasional dinosaur footprints. From cross-bedded sandstone

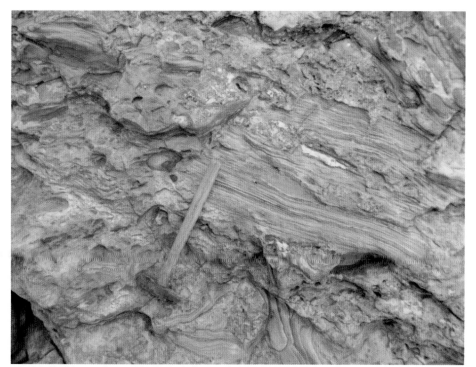

Figure 43. Rip-up clasts of laminated siltstone in Black's Channel 'E' at Cromer Point.

here Romano and Whyte (2013) found several trace fossils attributed to horse-shoe crabs (*Selenichnites*), suggesting marine-influenced conditions. Out on the rock platform the position of the meander belt is occupied by a series of almost level-bedded, ripple-marked sandstones alternating with drab shales containing plants (the Scalby Plant Bed). Here we have the floodplain fines, laid down at some distance from the stream, and not reworked as most meander-belt floodplain fines were.

2B. The rock platform north-east of Scalby Lodge. South of channel G, a meander-belt channel is seen in the rock platform, running a straight course with a predominant cross-bedding dip to the east and a subsidiary westerly dip. The dip direction of infilled trough cross-bedding shows a northerly current flow. Adjacent to this channel is a series of almost flat lying sandbanks, former crevasse splays. Adjacent to the small valley in the cliff, the sandbank nearest present high-water mark displays a 10 m long trail of bipedal dinosaur footprints (Fig. 44). Photographs of this locality were published by Delair and Sargeant (1985, fig. 3) and Whyte *et al.* (2010, fig. 11), and a drawing of the full trackway is given by Romano and Whyte (2003). The poorly sorted fluvial sand was compacted under and around the prints, which are now being revealed by selective erosion of the overlying argillaceous sandstone.

2C. Cowlam Hole. Opposite this southernmost of the two small valleys (Fig. 44), the meander-belt channel makes a spectacular 180 degree sweep across the rock platform (Fig. 45). The meander scrolls swing round from north-east through south-west to south-

3 Excursions: Itinerary 6

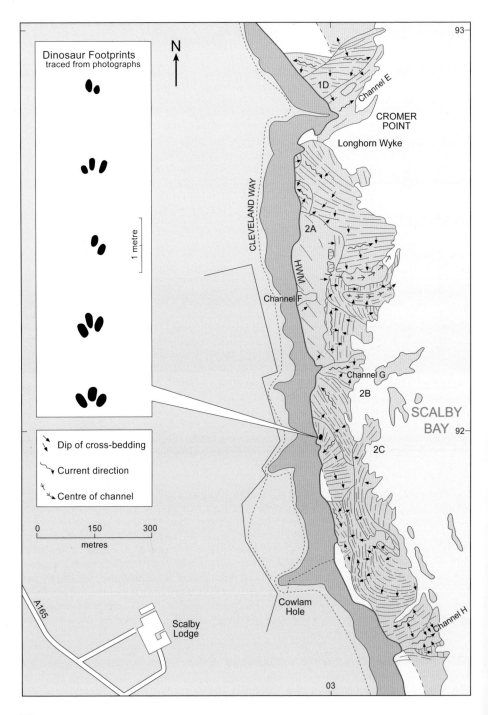

3 Excursions: Itinerary 6

Figure 44. (facing page) Map of meander-belt channels in the Scalby Formation, Scalby Bay, as exposed in the rock platform in Scalby Bay. Cliff in darker brown shading. Based on field mapping by J.K. Wright.

east. The dip direction of infilled trough cross-bedding shows that the current flow was clockwise. Seawards, a later gently curving channel cuts across the meander and also cuts across several earlier channels to the north. Near high-water mark at TA 0316 9152, Romano and Whyte (2003, fig. 26) noted the presence of both the pes and manus prints of a sauropod trackway preserved on a sandbar within the channel sandstones.

2D. Scalby Ness. If time allows, proceed now across Black's Channel H into the southern end of Scalby Bay and round onto Scalby Ness (Fig. 37). The gentle seaward dip carries down to sea level Long Nab Member shales which overly the meander belt. At the base of the cliff on the headland these consist of a cross-bedded, channel-fill sandstone which has clay drapes over the thinly bedded sand laminae. The clay drapes mark pauses in the current sufficient to allow plant debris to settle out. Fronds of *Ginkgo huttoni* are very common. Southwards, a series of small faults brings up the Moor Grit into the base of the cliff. As before, the top surface of the Moor Grit is rooted and deeply sideritised. It is now necessary to climb the footpath up the grassy cliffs here to gain access to the cliff path, from where it is a 2 km walk back along the Cleveland Way to Burniston car park.

Figure 45. Meander scrolls in meander belt, Scalby Bay.

ITINERARY 7: EGTON BRIDGE TO GROSMONT

S.E. Livera, P.F. Rawson and J.K. Wright

OS   1:25 000 Explorer OL 27 North York Moors Eastern area
           1:50 000 Landranger 94 Whitby & Esk Dale.
GS   1:50 000 Sheet 35 & 44 Whitby and Scalby

This itinerary focuses on both solid geology (Locs. 1, 5 & 6), including the Cleveland Dyke, and glacial features (Locs. 2–4) in an area that lay at the margin of the last glacial (Devensian) ice sheet. The maximum extent of ice during this glaciation (Fig. 6) shows that much of the North York Moors was ice-free at that time, experiencing periglacial conditions, but that the whole of the present coastal area was blocked by ice and moraine.

The glacial history of Eskdale and the adjacent northern part of the moors was studied by Kendall (1902) who recognised a series of ice-dammed lakes, including 'Lake Eskdale', linked by drainage ('overflow') channels. Subsequently, Gregory (1962, 1965) suggested that rather than being occupied by a lake, much of west and central Eskdale was under ice, and Straw (1979) recorded a maximum drift limit there of 245 m O.D.

According to Gregory (1962), the glacial drainage channel network in Eskdale appears to have originated through the development of subglacial streams and water flowing at the ice margin and there was a complex pattern of subsequent deglaciation. Although it is now accepted that there was no Lake Eskdale, he did confirm Kendall's (1902) view that a small lake occupied the Wheeldale valley during the early stages of deglaciation. This was fed by two sets of ice-marginal drainage channels carrying water southward into the lake and Gregory (1962) recognised two stages in the development of this drainage system. The first, higher stage, was the Lady Bridge Slack to Hollins channel system which formed the higher flat of Lake Wheeldale at about 198 m O.D. The later, lower stage (183 m) was fed through Moss Swang and Randay Mere. During this second stage the outlet of Lake Wheeldale was through the Goathland Church Channel towards Newtondale (Fig. 46) and thence to Lake Pickering (Fig. 6). This lower drainage system is the one which will be examined at Localities 2–4.

Although this itinerary is laid out as a circular tour, localities 1, 5 and 6 could be visited individually while the features of the Moss Swang glacial channel at localities 2–4 are best viewed in that order. The account of localities 2–4 is modified from that of Hemingway (1968) in an early edition of this guide. Unfortunately the type locality of the Middle Jurassic Eller Beck Formation at Eller Beck (Knox, 1973), which was described in all previous editions, is no longer safely accessible.

Locality 1. Duckscar Quarry, Egton Bridge (Cleveland Dyke)

Parking is available for several cars at this disused roadside quarry a short distance to the west of Egton Bridge at NZ 798 053. The quarry is overgrown but the face shows a well-exposed section across the Cleveland Dyke, which here is 9 m wide. The shaly country rock can be seen on both sides of the dyke and, adjacent to the tholeiitic basalt, it is metamorphosed to a hornfels for a few centimetres and in places is brecciated. The occurrence of *Pseudomytiloides dubius* in the country rock shows that here the dyke cut through the Mulgrave Shale Member of the Whitby Mudstone Formation.

3 Excursions: Itinerary 7

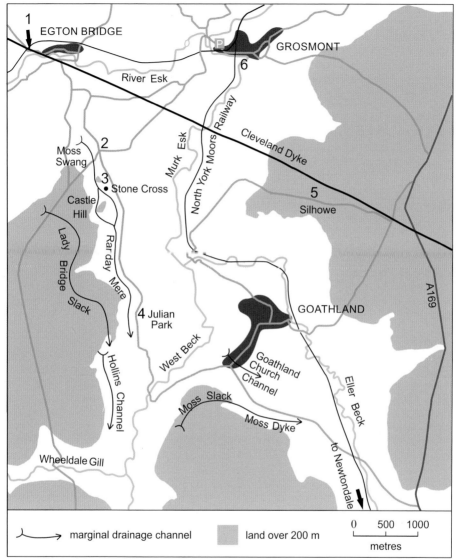

Figure 46. Map of the Egton Bridge, Goathland and Grosmont area (Itinerary 7).

Locality 2. Moss Swang (glacial features)

Return to Egton Bridge, turn right and follow the road signposted Goathland for 2 km to the crossroads at NZ 807 038. Pause here to observe the great, flat-floored channel of Moss Swang just west of the road. This is one of the finest of the marginal glacial drainage channels which carried water to the south. Here, both sides of the channel are cut in solid rock. Turning to the east, observe the view down lower Eskdale and across the Eskdale Dome. This structure yielded natural gas from which Whitby and district was supplied in the 1960s. Notice in particular the flat-topped feature at approximately 76 m O.D. which extends along the valley side near Newbiggin Hall [NZ 841 069] for nearly 2 km. This surface, cut in glacial till, may be traced at rather higher levels above Grosmont and is a high terrace of the late-glacial Esk.

Locality 3. Stone Cross and Castle Hill (glacial channels)

About 400 m south of the crossroads park on the roadside at NZ 808 035, where a public footpath sign and stile can be seen on the west side of the road, and walk to a nearby monument, Stone Cross. To the south of this viewpoint, across an intervening valley, is Castle Hill. The valley is part of the channel system transporting meltwater to Lake Wheeldale and forms the main Moss Swang channel, which starts at locality 2 and passes round the west side of the hill on which the monument stands before swinging eastward in front of Castle Hill then turning southward again. To the west of Castle Hill a higher level channel can clearly be seen, cutting down through the solid rock for about 15 m and hanging at its northern end about 15 m above the main channel. Kendall (1902) thought this feature indicated the readvance of a small lobe of the ice front over the older Moss Swang channel so that meltwaters were forced to cut a new channel which reached a depth of only 15 m before a recurrence of retreat caused its abandonment. Gregory (1962), however, believed that the main channel and the hanging channel developed simultaneously until they attained a common floor level during a period when the ice margin was stable. Then as the ice retreated and the amount of water flowing through the main channel diminished, the higher level itinerary was abandoned.

Locality 4. Julian Park (glacial features)

Continue southward to Julian Park [NZ 814 010] where the channel lost its identity as the waters emptied into Wheeldale Lake. The sediments of its flat floor, built up to approximately 156 m O.D., completely choked the pre-existing river valley so that post-glacial drainage followed a new line cutting the rock gorge at New Wath Scar [NZ 820 006] and Hollins Wood.

From Julian Park continue eastward through Goathland onto the minor road crossing NNE over Goathland Moor. At NZ 852 028 take a left fork and follow the road for 1.2 km.

Locality 5. Silhowe (Cleveland Dyke workings)

Park at the side of the road at NZ 840 031 where a track leads northward along an old tramway for 250 m to a deep, elongate cut where the Cleveland Dyke was formerly quarried (Fig. 47). The cutting here gives a good impression of the scale of working. Although all the basalt has been removed, occasional loose fragments can be found and the country rock is well exposed showing baked margins. The country rock here is the Scalby Formation, consisting of thin sandstones, occasionally thickening up into small sand-filled channels, interbedded with dark grey shales.

Returning to the road follow another track and cross open moor southwards for about 400 m to the remains of a basalt mine which is marked by a ruined workshop and arched adit entrance [NZ 840 027]. The adit (closed) runs 540 m to the north-east to intersect the dyke 60 m below the ground surface. The mine was worked from 1899 to 1950 and the extensive underground mine supplied rock via a tramway to a crushing plant at Goathland station from where it was exported by train.

3 Excursions: Itinerary 7

Figure 47. Former workings in the Cleveland Dyke, Silhowe. Channel-fill sandstones in the Scalby Formation are visible in the left foreground and in the far distance to the right of the cutting.

From here return eastward to the road junction, turn left and continue for 700 m to turn left onto the A169, continue for about 500 m then turn left again onto a minor road signposted for Grosmont. Continue for 4 km to the village car park on the west side of Grosmont.

Locality 6. Murk Esk (Cleveland Ironstone Formation)

From the car park walk into the village, cross the railway and immediately turn right onto a path leading over a footbridge towards the old railway tunnel. Walk up to St Matthew's church, go around the back of the church through the churchyard and turn left onto a track towards the river. After about 75 m turn right onto a public footpath to cross the Murk Esk using the footbridge at NZ 829 051. The footings of this bridge are set on the Two Foot Seam of the Cleveland Ironstone Formation and the cut bank on the western side of the river exposes the overlying Pecten Seam. Walk north along the footpath next to the river (but note that the path is impassable after heavy rain) to NZ 830 052 where there is a good exposure of the sequence from the Two Foot Seam to the paper shales forming the lowest part of the Grey Shales Member of the Whitby Mudstone Formation. The exposure becomes overhung and dangerous to the west and extreme caution should be taken if walking in this direction. The section shows the two beds (or lifts in mining terminology) of the Pecten Seam and a further three ironstone seams above, which are the thinned, inland and southerly, equivalent of the Main Seam worked in the Guisborough to Staithes area. The lower bed of the Pecten Seam contains abundant large *Pseudopecten inequivalvis* that can be seen in fallen blocks in the river. The seams are muddy, bioturbated, shelly and ooidal sediments with a locally pervasive siderite cement, most notably developed at the top of the Two Foot Seam.

85

Grosmont developed as a settlement around the railway and ironstone mines which were worked and processed here from 1836 to 1891, starting well before the larger mines to the north. Two main seams were worked in the Esk Valley, the lower Avicula Seam (not exposed) and the overlying Pecten Seam, both being of similar thickness. The iron content averaged around 26% Fe and a large part of the area is undermined by bord and pillar workings. An adit entrance can be seen on the southern bank of the Murk Esk from the footbridge near the level crossing in the village centre. There are detailed descriptions of the industry in Chapman (2002) and Goldring (2006).

ITINERARY 8: SOUTH BAY, SCARBOROUGH AND CORNELIAN BAY

J.K. Wright

OS   1:25 000 Explorer 301 Scarborough, Bridlington & Flamborough Head
           1:50 000 Landranger 101 Scarborough
GS   1:50 000 Sheet 54 Scarborough

This itinerary demonstrates Middle Jurassic sequences south of Scarborough (Fig. 48). The 2 km route follows the beach and rock platform through South Bay, Scarborough into Cornelian Bay (Fig. 49). The going is fairly straightforward. The excursion is best done on a falling tide. Difficulties may be experienced in several places when the tide is in, though there is no danger of being cut off. In particular, the sea reaches right up to the base of the cliff at Locality 3A and one should not rely on being able to pass this point within three hours of high tide.

It is possible to start the itinerary at either South Bay or Cornelian Bay and the Rotunda Museum (Loc. 1) could provide a starting or a finishing point depending on the tides. The nearest car park to the Rotunda is an underground car park at the foot of Valley Road; there is also nearby street parking and other car parks are within 5 minutes' walk. Alternatively, the park-and-ride services from Seamer Road and Filey Road stop opposite the Rotunda. It is also possible to join the itinerary at Locality 3 by parking in the Holbeck Car Park [TA 049 868—maximum stay 3 hours] and walking down the adjacent path over the landscaped landslide surface. At Cornelian Bay there is a small car park at TA 057 861 (accessed via Cornelian Drive): from there pass through a gate, turn left on the nearby footpath (Cleveland Way), walk down for about 50 m then leave the path to join a wide track leading steeply downwards, then join a narrow path to the right leading down to the shore.

Locality 1. Rotunda—The William Smith Museum of Geology

In 1829 the recently formed Scarborough Philosophical Society opened a museum which holds the unique distinction of having been designed under the guidance of William Smith to illustrate his geological principles in a purpose built, circular gallery. At the time, Smith was employed as a land agent to Sir John Johnstone, a Fellow of the Geological Society, at Hackness Hall. Johnstone donated the stone, Hackness Building Stone, quarried on his estate. Smith also acted as site manager during the building and remained associated with it until his death in 1839.

The wings, also of Hackness Building Stone, were added in 1860. Eventually, the Rotunda was reserved for archaeological displays and the geological specimens, having been kept in storage for many years, were transferred to the Woodend Natural History Museum. Following the closure of Woodend, a major fund-raising campaign led by Lord Derwent, the direct descendant of Johnstone, and Scarborough Borough Council, resulted in the Rotunda being extended and reopened as 'Rotunda—the William Smith Museum of Geology' to provide a fitting tribute to Smith and form a geological museum of national and international historical significance (Fig. 50). It is now managed by Scarborough Museums Trust (SMT).

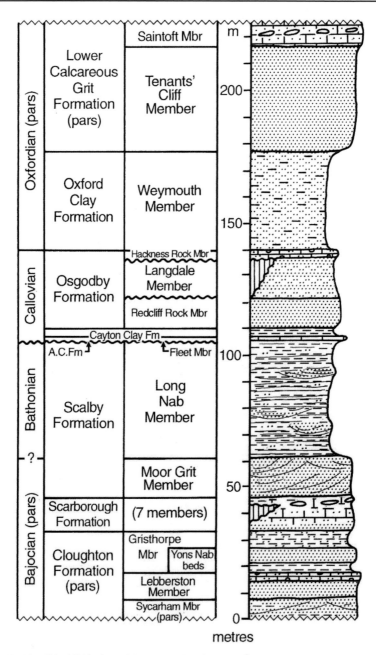

Figure 48. Simplified lithic log of the succession from South Bay to Cayton Bay. For lithological key see Fig. 15. From Rawson and Wright (1996, fig. 13). Reproduced by permission of the Geological Society of London.

3 Excursions: Itinerary 8

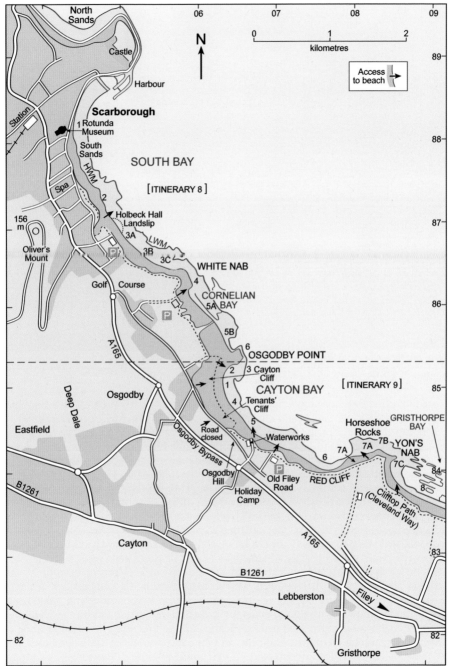

Figure 49. Map of localities in the Scarborough to Yons Nab area (Itineraries 8 and 9). Cliff in darker brown shading.

3 Excursions: Itinerary 8

Figure 50. Rotunda—the William Smith Museum of Geology, Scarborough.

Locality 2. Cliffs south-east of The Spa (Scalby Formation)

Proceed from the Rotunda along the Spa approach and past Scarborough's Victorian Spa. Below the Clock Café and beach chalets 200 m south of the Spa, 5 m high cliffs in the Moor Grit Member of the Scalby Formation are available for study. The Moor Grit was laid down after an uplift of the Mid North Sea High led to erosion of source areas and river systems carrying large amounts of sandy sediment sweeping down onto the Cleveland Basin alluvial plain. At the base of the first exposure, cross-bedded sand/silt alternations formed on a levee adjacent to a channel are cut through by massive sand in a crevasse splay. These beds are overlain by 0.5 m of fine, level-bedded sandstones containing numerous resting traces of the bivalve *Unio*, known as *Lockeia*. These sandstones are clearly lacustrine in nature, accumulating in a lake situated on the alluvial plain in between distributary channels. They are followed by 2.5 m of largely level-bedded sandstones with small, slumped channel structures (Fig. 51), these beds being without trace fossils but with probable dinosaur footprints on the top surface. The *Lockeia* beds and overlying sandstones can be followed for 100 m eastwards. Cross-bedded sandstones become more prominently developed above until a cross-bedded unit with a persistent easterly dip laid down in a migrating channel cuts right down and erodes the *Unio* bed. The individual cross-bedded sand beds are lenticular, each marking the rapid infilling of a channel cut through the underlying beds, and the sand overlain by thin, laminated silts and muds as the water flow ameliorated. At one point, collapse of the bank of a channel choked with sand has led to slumping of channel sand contorting the adjacent laminated silts and shales (Fig. 51).

3 Excursions: Itinerary 8

Figure 51. Sedimentary structures in the Moor Grit Member, near the Spa, South Bay. Slumping of channel sand (behind 30 cm hammer) has contorted the underlying laminated offbank deposits.

Locality 3. South Bay, Scarborough (Scarborough and Scalby formations)

Just south of Locality 2, in July 1993, the upper cliff collapsed right across the rock platform in a gigantic landslip, taking the Holbeck Hall Hotel with it. The landslip has now been stabilized, grassed over and the toe protected by a supporting fringe of boulders of Scandinavian granite. Proceed along the gravel track leading across the landslip and carefully down the concrete ramp onto the shore. The rock platform consists of the tough, sideritic White Nab Ironstone Member of the Scarborough Formation. Overlying the ironstone is shelly, argillaceous limestone. Bioturbation is marked with networks of burrows (*Thalassinoides*) in the ironstone infilled with shelly, argillaceous limestone from the overlying bed. Sideritic concretions are developed at several horizons. Bivalves are abundant, particularly *Gervillella*, *Meleagrinella* and *Pleuromya*.

3A. Cliffs and rock platform south of the landslip. Just above the base of the cliff, there is a thin representative of the fine-grained, bioturbated sandstone of the Bogmire Gill Member of the Scarborough Formation, succeeded by the massive, coarse-grained, cross-laminated sandstone of the Moor Grit. Erosion of the Bogmire Gill Member beneath the Moor Grit is very evident. Scour channels infilled with clay clasts, charcoal fragments and fossil wood cut down into the Bogmire Gill Member. The Moor Grit comprises here a 6 m sequence of north-westerly dipping cross-bedded units. These were laid down as point-bar deposits as the channel carrying the sediment gradually migrated north-westwards. Frequent minor channels cutting through these deposits have left small trough cross-stratified infills with bedding dipping in the opposite direction to the main cross-bedding. Ravenne (2002) and Hesselbo *et al.* (2003) consider that the Moor Grit of South Bay is not contiguous with that seen to the north (Itinerary 5), but forms a separate 'Prism' II valley fill, overlying the valley fill of Prism I to the north. The two Prisms are nowhere seen in contact, however, and Hesselbo *et al.* even place Prism II below Prism I.

Cliff falls here occasionally reveal well-preserved dinosaur footprints.

3B. The centre of South Bay. One hundred and fifty metres south of the landslip, the Scarborough Formation is overlain by cross-bedded channel sandstones dipping southwards (Fig. 52). With favourable beach conditions, scouring of the uppermost Scarborough Formation can be demonstrated; the basal part of the Moor Grit being coarse and ill-sorted. The persistent depositional dip in the lower cliff can be followed for several hundred metres southwards. Frequent small channels are filled with trough cross-stratified sandstone. The higher Moor Grit demonstrates deposition is a much less stable environment. Many channels are seen in section in the cliff (Fig. 52): most were infilled with sand and aggraded rapidly with little migration of the channel. Some channels were infilled with laminated silt and clay and others became stagnant oxbow lakes infilled with plant debris and clay. The upper Moor Grit thus formed in an unstable environment. Periodic floods eroded the underlying beds and the stream channels aggraded rapidly as they became choked with sediment. Crevasse-splaying produced sheet sands alternating with alluvial silts and clays, and the abandoned channels were filled with fine-grained sediment.

3C. The south end of South Bay. The southward migration of the channel laying down the persistent, southerly dipping, cross-bedded sandstone unit can be followed towards the southern end of the bay, where these sands pass laterally into abandoned channel fa-

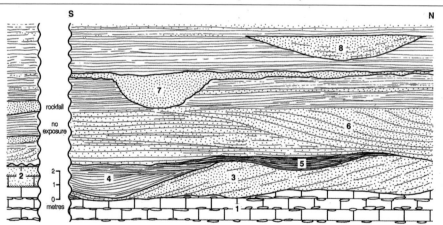

Figure 52. The Scarborough Formation (1, 2) and the Moor Grit and Long Nab Members (3–8) in South Bay. A schematic diagram to illustrate the main features seen in the cliffs in the southern part of South Bay; not all these features are seen in any one section. (1) White Nab Member. (2) Bogmire Gill Member (very-fine-grained marine sandstone). (3) Epsilon cross-bedded sands laid down by a mature river channel migrating steadily southwards. (4) Channel infill of silt, clay and fine sand. (5) Abandoned channel infilled with carbonaceous debris. (6) Rapidly aggrading channel migrating northwards. (7) Crevasse-splay channel rapidly eroded and infilled with unbedded sand. (8) Short-lived channel crossing the alluvial flats.

cies, with alternations of laminated, argillaceous siltstone and sandstone (Fig. 52). South of this, erosion beneath the Scalby Formation has proceeded to a higher base level than to the north, and the interval between the White Nab Ironstone and Moor Grit Members is occupied by the quasi-marine beds poorly seen at Locality 1A. This is the Bogmire Gill Member of the Scarborough Formation (Gowland & Riding, 1991). Best seen at the far end of the sandy beach, the grey, shelly limestone described above is overlain by 2.1 m of white, level-bedded, fine-grained sandstone showing delicate scour-and-fill cross-bedding. At three horizons bioturbation is well marked. Some large pedestals on the upper rock platform show beautifully the erosion of this fine-grained sandstone, with hollows cut in its top surface infilled with the coarse sand of the Moor Grit. The Bogmire Gill Member was clearly well indurated before being eroded by the Moor Grit streams.

Eighty metres southwards, in the basal Moor Grit, is a bed of sideritic mudstone laid down in a small, abandoned channel, and from here the remains of the fish *Heterolepidotus* and turtle plates have been found (Scarborough Collections, managed by the Scarborough Museums Trust). The sloping upper cliff is formed of the Long Nab Member, 45 m thick, which consists chiefly of non-marine shales with occasional channel sands. These beds occupy the cliff up to the Abbotsbury Cornbrash Formation exposed just beneath the Golf Course. Many blocks of Fleet Member limestone containing abundant *Trigonia elongata* and *Myophorella scarburgensis* can be found on the upper beach in the centre of the bay. Occasional ammonites (*Macrocephalites* sp.) can also be found in these blocks.

Locality 4. White Nab (Scarborough and Scalby formations)

At White Nab, a gentle anticlinal structure has brought up the Scarborough Formation sufficiently to enable a 6.5 m succession to be examined working out towards low-water mark. The White Nab Ironstone Member comprises a 2.5 m succession of alternations of sideritic mudstone and calcareous, fossiliferous shale (Parsons, 1977). The best preserved fossils, including ammonites, occur in concretions in the sandy shales. The attribution of the beds beneath has proved difficult, as there is no obvious correlation with other nearby successions in the Scarborough Formation. A massive, 2 m thick sandy limestone containing *Pseudomelania* was quarried here in Victorian times as building stone for the pier at Scarborough. It may represent the Lambfold Hill Grit Member (Parsons, 1977). Shales with nodules and a band of shelly mudstone seen to a thickness of 1 m near low-water mark have yielded *Stephanoceras*. This ammonite fauna is too young for these beds to be included in the Ravenscar Shale Member.

In the base of the cliff at White Nab, the White Nab Ironstone is succeeded by 90 cm of shelly, bioturbated, delicately laminated siltstones of the Bogmire Gill Member. These are followed by 3 m of gently cross-bedded fluvial sandstones and then by strongly cross-bedded sandstone. The White Nab succession thus shows a steady, step-by-step progression from marine limestones through to strongly cross-bedded fluvial sandstones.

Locality 5. Cornelian Bay (Scalby and Osgodby formations)

On the south side of White Nab, the Moor Grit strikes east-west across the rock platform as the beds take on a southerly dip. Climb over the Moor Grit outcrop into Cornelian Bay (Fig. 53), noting the beautifully displayed cross-bedding on the south side of the outcrop. The top surface of the Moor Grit here reveals poorly preserved dinosaur footprints. The going is much more straightforward now across the sandy beaches of this pleasant little bay, more remote and less frequented by holidaymakers.

Figure 53. View looking SE along Cornelian Bay to Osgodby Point: Red Cliff and the Red Cliff Fault, Cayton Bay (see Fig. 59), are visible in the distance behind the point. The channel capped by a pillbox lies at the centre of the photograph.

5A. The centre of Cornelian Bay. Along most of Cornelian Bay the beds are almost horizontal, with the Long Nab Member of the Scalby Formation exposed in the low cliff, though sometimes obscured by slipped glacial till. The member consists predominantly of shale when first seen, with thin sheets of sandstone developed at several levels, and occasional cross sections through sand-filled channels. In the centre of the bay, a prominent channel sandstone, capped by a wartime pillbox, cuts across the beach (see Fig. 53). Here, the relations between channel sandstones, siltstones and clays can be demonstrated in the cliff section. The clays pass laterally into alternations of clay with thin siltstones laid down during periodic flooding by the adjacent channel. Close to the channel, several crevasse-splay sandstones each cut across laminated silts and muds, which are highly contorted. The reasons for this have caused some debate. Romano & Whyte (2003) suggested that the contortions are due to dinosaurs walking over the soft sediment, churning it up ('dinoturbation'). The alternative view is that the distortions are formed by water escape caused by the sudden influx of sand. The lowest, very carbonaceous bed here is almost certainly a palaeosol. The planar-bedded sandstone at the top then passes laterally into the southerly dipping channel sandstone. This channel cuts right down onto the Moor Grit out on the rock platform, where there are lenses in the Moor Grit packed with fossil charcoal derived from a forest fire upstream.

On top of the channel, Romano and Whyte (2003, fig. 10) illustrated sauropod footprints. Just above the channel and visible in the low cliffs either side is a very interesting lenticular bed of fine-grained sand. Every 2–3 m, concretions are developed at its base, descending into the underlying muds. Romano and Whyte (2003, figs 7, 18) interpreted these as infilled sauropod footprints. Proceeding southwards, after a 100 m gap due to slipped till, a 2 m high cliff is reached. Massive sandstone (60 cm) is seen resting on 1 m of mudstone with intensely contorted beds of sand. Up to 5 incursions of sand are seen, each very deformed and injected into the underlying mud. These structures have again been interpreted to represent either dinoturbation or water escape following rapid incursions of sediment.

5B. The southern end of Cornelian Bay. Some 150 m before the southern end of Cornelian Bay the western branch of the Cayton Bay Fault is crossed. The downthrow of 45 m to the east brings the Abbotsbury Cornbrash Formation down to the beach (note: the Cornbrash formed the cliff top at loc. 3C). It is exposed just below high-water mark when the base of the cliff has been cleared of boulder clay by storms, but it can be completely covered by beach gravel in summer. All four subdivisions of the Cornbrash (Wright, 1977) are present, with occasional *Macrocephalites* sp. There then follows the type succession of the Osgodby Formation. The three members of the formation are well developed here: the Red Cliff Rock Member, 3.5 m of iron rich sandstone containing nests of fossils including numerous bivalves and *Kepplerites*, the Langdale Member (3.8 m), bioturbated sandstone with moulds of belemnites and *Gryphaea* overlain by bioturbated silts with *Erymnoceras*, and finally the Hackness Rock Member (0.77 m), a sandy, chamositic limestone containing *Quenstedtoceras* and *Kosmoceras*. Much of this sequence can be covered by slipped glacial till after heavy rain. The overlying Oxford Clay is well displayed, and this locality was at one time considered to be the type locality for the Callovian/Oxfordian junction. The basal Scarburgense Subzone of the Oxfordian is thin (35 cm), but yields phosphatised *Cardioceras scarburgense*. The overlying thick, silty clays of the Weymouth Member of the Oxford Clay have yielded *Cardioceras praecordatum* and *Peltoceras* (Wright, 1968a).

Locality 6. Osgodby Point (Millepore Bed, Lebberston Member)

The eastern Branch of the Cayton Bay Fault, with an upthrow of 110 m to the east, cuts through the headland of Osgodby Point and brings the Millepore Bed up to form a natural barrier protecting the headland. In fact, the Eller Beck Formation is brought up along the line of the fault in the centre of Cornelian Bay, but this exposure is only visible at very low tide (Livera & Leeder, 1981). The fault plane is exposed in the rock platform, Hackness Rock and Oxford Clay being faulted against poorly sorted, micaceous sandstone of the Sycarham Member. The fault plane dips at 45° to the WSW. The wartime pillbox sits upon the lowest beds of the Millepore Bed, which consists of very sandy, shelly ooidal limestone, sometimes an ooidal calcareous sandstone. The cross-bedding, ooids and the broken nature of the shell fragments, all point to deposition of this bed in a shallow, intertidal, high-energy but fully marine environment. The Millepore Bed limestones are crisscrossed by numerous calcite-filled veins running at 45 degrees either side of the fault direction and formed during the compressive stress regime created by the Alpine orogeny.

The route back to Scarborough is to proceed to the north end of Cornelian Bay, from where steps and a broad track lead up the cliff to the coastal path (Cleveland Way) leading to Scarborough.

ITINERARY 9: CAYTON BAY, YONS NAB AND GRISTHORPE BAY

J.K. Wright

OS   1:25 000 Explorer 301 Scarborough, Bridlington & Flamborough Head
         1:50 000 Landranger 101 Scarborough
GS   1:50 000 Sheet 54 Scarborough

This itinerary demonstrates much of the Ravenscar Group, with the non-marine or marginal-marine sequences of the Gristhorpe Member and the Scalby Formation, and the marine intercalations of the Lebberston Member and the Scarborough Formation. Also well exposed are the fully marine Middle and Upper Jurassic sequences of the Abbotsbury Cornbrash, Osgodby, Oxford Clay and Lower Calcareous Grit formations. Roadside parking is normally available on Osgodby Hill [TA 065 842] (Fig. 49). Alternatively, cars can be parked at the nearby Cayton Bay Surf School pay car park off Old Filey Road [TA 068 842]. A falling tide is necessary.

Stratigraphically, the itinerary is less straightforward than previous itineraries. Several faults belonging to the Peak–Red Cliff Fault Zone (Peak Trough) run obliquely into the coastal area from Castle Hill southwards. As one crosses and recrosses the faults one jumps up or down the succession several times, making reference to the lithic log (Fig. 48) necessary. The advantage is that one is able to see a substantial part of the N.E. Yorkshire Middle and Upper Jurassic successions within a short distance. Glacial till regularly slips down over parts of the lower cliff and this means that some sections described here may not be accessible during a particular visit.

Cross the style in Osgodby Hill at [TA 065 843] and proceed across the field to join the Cleveland Way at the signpost (alternatively, from the car park, follow the Cleveland Way to this point). Follow the Cleveland Way down the steps and north through the system of small valleys which is a large, stabilized area of landslipped blocks of Lower Calcareous Grit, to be discussed later. Follow the path up and down the blocks till it reaches the cliff edge with an excellent view of the north end of Cayton Bay (Fig. 54). Opposite is Osgodby Point, tipped by Gristhorpe Member sandstone, with an outer protective reef of Millepore Bed. A fault, the main Osgodby Fault (Fig. 54) separates these beds from the Osgodby and Oxford Clay formations dipping north in the cliff. In the foreground, in the rock platform are level-bedded reefs of Lower Calcareous Grit.

Locality 1. Cayton Cliff landslide

Proceed through the wood and at the path junction turn right down the steps to the beach. A large area of boulders of shelly, ferruginous sandstone derived from the Red Cliff Rock, and of slipped, plastic clay (Cayton Clay Formation) has yielded interesting ammonites (*Kepplerites* sp., *Pseudocadoceras boreale*, *Macrocephalites* sp.).

The wartime pillbox sits on the toe of the Cayton Cliff landslide (Johnson & Fish, 2012). Heavy rain early in 2008 resulted in reactivation of this very unstable area underlain by glacial till and large toes of landslide debris swept across the beach. Buildings and areas of woodland were brought down and tree stumps can still be seen rising from the beach.

Figure 54. (facing page) Geological map of the strata exposed in the rock platform in Cornelian and Cayton bays. Cliff in darker brown shading. After Wright (1978, fig. 2.6), with additional data from Johnson and Fish (2012).

3 Excursions: Itinerary 9

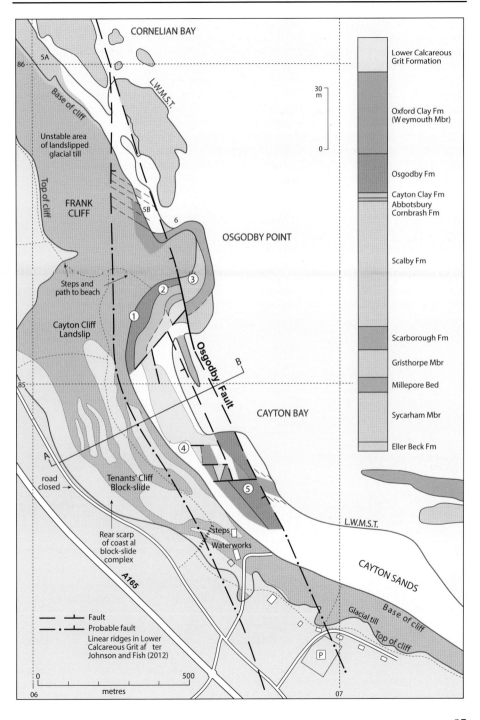

Locality 2. South side of Osgodby Point, Osgodby (Cayton Clay and Abbotsbury Cornbrash formations)

Towards Osgodby Point, a large area of sandy beach is underlain by the Cayton Clay and Abbotsbury Cornbrash formations and these beds are usually revealed after winter storms. The Cornbrash forms the reefs at mid-tide level. It is fossiliferous, with bivalves and *Macrocephalites kamptus*. Cut, laquered blocks show excellent bioturbation structures. The Cayton Clay yields phosphatic nodules containing the lobster *Glyphea*, and *Macrocephalites kamptus*, with authigenic sphalerite. Exposures of Red Cliff Rock and Hackness Rock can be found amidst the boulders. The Hackness Rock yields interesting perisphinctids and *Quenstedtoceras* sp. The beds dip steeply north-west into the cliff and are separated from the cliff by a small strike fault.

Locality 3. Osgodby Point (Gristhorpe Member, Lebberston Member (Yons Nab Beds and Millepore Bed))

As one approaches Osgodby Point, faulted against the Oxford Clay is the Gristhorpe Member, forming a cliff consisting of a 4 m thick sequence of cross-bedded sandstone displaying herringbone structure (Fig. 55), typical of beds laid down in estuarine conditions, overlying 2 m of thinly bedded sands and silts showing ripple-drift lamination and flaser bedding. These latter beds can be regarded as quasi-marine beds belonging to

Figure 55. Herringbone cross-stratification in the Gristhorpe Member at Osgodby Point. Beneath are level-bedded, marginally marine Yons Nab beds.

the Yons Nab Beds. Bate (1959) suggested that the cross-bedded sands also belong to the Yons Nab Beds, but similar cross-bedded sandstones at Yons Nab (see below) were placed in the Gristhorpe Member by Livera and Leeder (1981).

The area of Osgodby Point is littered with blocks of cross-bedded sandstone and of the tough, shelly limestone of the Millepore Bed. The limestone can be examined in great detail in the blocks and in the intersecting, weathered joint surfaces of the pedestals of limestone present here (Fig. 56). It consists of cross-bedded, shelly ooidal limestone, there being at least 5 courses of limestone, the cross-bedding direction in each course varying from one to another, but also varying within courses. The two highest beds have in their top surfaces extensive infilled *Thalassinoides* networks, suggestive of pauses in sedimentation, but the lower courses of limestone, which are more ooidal (Itinerary 8) show no such evidence. Bedding surfaces in the more shelly blocks can be strewn with *Pentacrinites* ossicles, and there are numerous *Haploecia* [*Millepora*] *straminea*, and this exposure can be regarded as the type locality of the Millepore Bed. It was laid down in very shallow water, but under fully marine conditions, with strong tidal currents sweeping the coarse shell sand and ooids and producing the strong cross-bedding. The limestone is then overlain by ferruginous, fine-grained sand with plant remains, the lowest bed of an 8 m thick sequence of marginal marine, bioturbated, almost level-bedded sandy clays and clayey sands representing the Yons Nab Beds. These are largely removed by the downcutting of the channel sand at the southern end of the point.

Figure 56. The Millepore Bed at Osgodby Point, showing shallow marine cross-bedding. The outcrop is broken by strong N-S and E-W trending joints.

Locality 4. Tenants' Cliff (Lower Calcareous Grit and Tenants' Cliff landslip)

Proceed now over the sandy beach towards the landslipped area of Tenants' Cliff. At the foot of the grassy cliffs there is an extensive 300 m x 100 m area of flat-lying Lower Calcareous Grit in the rock platform (Fig. 54 – note: these strata are not marked on the Geological Survey map 54 (Scarborough), which is erroneous here). The flat-lying strata in the beach cannot be connected with the landslip, which consists of the large, slipped, rotated blocks of Lower Calcareous Grit and Oxford Clay in the cliffs above. The strata in the beach are *in situ* in a fault trough (Fig. 57), separated by faults from Ravenscar Group strata to the east and the full Lower Calcareous Grit, Oxford Clay and Osgodby Formation sequence present in Osgodby Cliff to the west. Once coastal erosion of these beds in the trough had proceeded sufficiently to reveal the western fault scarp (Fig. 57),

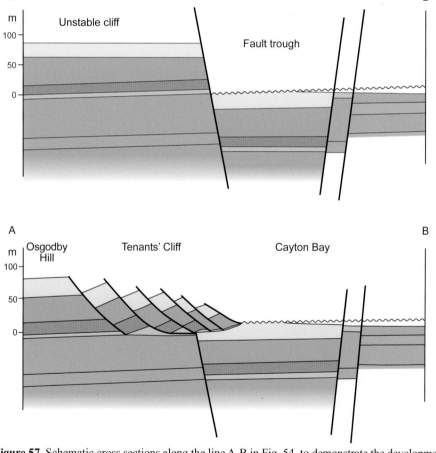

Figure 57. Schematic cross sections along the line A-B in Fig. 54, to demonstrate the development of the Tenant's Cliff block slide. Upper section, prior to block sliding, with coastal erosion reaching the proposed western fault in Fig. 54. Lower section, block sliding of Lower Calcareous Grit, silty Oxford Clay and Osgodby Formation along the slip plane provided by the Cayton Clay.

the cliff of Osgodby Formation, Oxford Clay and Lower Calcareous Grit was no longer supported and block sliding, apparently along the plastic clays of the Cayton Clay Formation, has carried huge blocks of these strata out over the fault trough. This interpretation is substantially one of those considered by Fish *et al.* (2006, fig. 4a), but not the one that they favoured: they preferred slipping along the Oxford Clay. However, this unit here is a tough, sandy, clayey siltstone (Wright, 1983, p. 255), which is still attached to the Lower Calcareous Grit in the rotated blocks. The junction of the Osgodby Formation sands resting on the plastic clays of the Cayton Clay Formation is a level where block sliding is much more likely to have begun.

The Tenants' Cliff Member of the Lower Calcareous Grit Formation is well seen both in the rock platform and in the fallen blocks below the landslip. It comprises a thick-bedded, fossiliferous calcareous sandstone. Almost all the fauna has been collected from the numerous concretions present here. Since the site was discovered early last century collectors have broken open the concretions to obtain the excellently preserved ammonite fauna of cardioceratids, perisphinctids and opeliids with additionally bivalves, brachiopods and gastropods. It is rarely worth attempting to break open a concretion unless cross sections of fossils are visible on the outside, as 90 per cent of concretions are barren.

Locality 5. Cayton Bay Waterworks (Scarborough, Osgodby and Oxford Clay formations)

In the large area of boulders opposite the Waterworks, at low tide a section in the Osgodby Formation and the Oxford Clay is seen dipping gently landwards (Fig. 54). These strata sit in a faulted trough, separated to the east by the Osgodby Fault from the Scarborough Formation, exposed at very low tide. When beach conditions are good there can be seen in these outermost reefs 5 m of fine-grained, argillaceous limestone and calcareous sandstone containing occasional bivalves, overlain by 1 m of massive, medium-grained shelly limestone. These clearly represent the Hundale Sandstone and Spindle Thorne Limestone members. Their presence here is very significant, for they are not present only 1.5 km away at Yons Nab, almost certainly due to contemporary movement of the Red Cliff Fault (see Loc. 7 below). Moving towards the waterworks, crossing into the fault trough, the next beds seen are the Red Cliff Rock Member, yielding very occasional ammonites, and a thin remnant of the Langdale Member, yielding only bivalves. The Hackness Rock Member, though covered by algae and seaweed in summer, yields numerous *Quenstedtoceras* spp. and *Peltoceras* spp. The Oxford Clay is frequently hidden by beach sand, but pyritised *Cardioceras* and *Peltoceras* can be collected.

Locality 6. Red Cliff (Abbotsbury Cornbrash to Lower Calcareous Grit formations)

Red Cliff is an imposing sight as it is approached across the wide, sandy beach (Figs 58, 59). The Osgodby Formation sandstones form the lowest quarter of the cliff, the steep slopes above are formed of the Oxford Clay Formation, while the Lower Calcareous Grit forms the vertical face in the upper part. Beginning at the base of the succession, the very fossiliferous limestones of the Abbotsbury Cornbrash Formation form a reef across the shore beneath Red Cliff [TA 0765 8405]. A thin limestone with numerous *Lopha marshii* overlies 37 cm of chamosite oolite limestone. *Rhizocorallium* from the base of the latter penetrate the underlying Scalby Formation siltstones (Wright, 1977) and hint at the major non-sequence which underlies the Cornbrash here. The lower Berry Member of the Abbotsbury Cornbrash Formation is absent throughout Yorkshire and the Fleet Member, as

3 Excursions: Itinerary 9

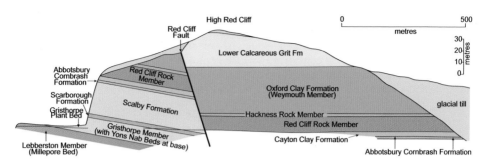

Figure 58. Cliff section at the south end of Cayton Bay.

Figure 59. Red Cliff and the Red Cliff Fault, Cayton Bay. The Red Cliff Fault forms the gulley in the middle of the picture. To the right is Red Cliff: the vertical face in the lower part of the cliff is formed by the Osgodby Formation, the slopes above expose the silty Oxford Clay Formation, while the top part of the cliff is formed by the Lower Calcareous Grit Formation. In the area to the left of the fault the Scarborough and overlying Scalby formations are exposed in the lower cliff face while the Osgodby and Oxford Clay formations form the highest part of the cliffs (see also Fig. 58).

seen here, is incomplete, 5 m of beds seen in inland sections being overstepped by the chamosite oolite limestone. Both the Fleet Member and the overlying Cayton Clay Formation are sometimes exposed at the NW end of the cliff [TA 074 842].

At the foot of Red Cliff the Osgodby Formation is accessible in places, though falling shale from above makes it dangerous to work there. The Red Cliff Rock Member makes up the major part of the formation and Red Cliff is the type section (Page, 1989). The bulk of the member is very sparsely fossiliferous and consists of intensely bioturbated, clean yellow sands with occasional very large calcareous concretions. Many fallen blocks of fine-grained chamosite oolite sandstone from the upper Red Cliff Rock Member occur on the upper beach and yield abundant bivalves and occasional ammonites, especially *Kepplerites*. The Langdale Member is absent here due to intraformational erosion (Wright, 1968a). The Hackness Rock Member comprises 1–2 m of chamosite oolite limestone at the top of the Osgodby Formation immediately beneath the Oxford Clay. It has yielded occasional *Quenstedtoceras* and *Kosmoceras*. Fallen blocks of Oxford Clay are occasionally fossiliferous with ammonites, but the blocks of Lower Calcareous Grit which litter the shore, though often riddled with *Thalassinoides*, are otherwise largely barren. This is surprising considering that Tenants' Cliff with its prolific fauna is only 1 km away.

Locality 7. Yons Nab (Ravenscar Group and Red Cliff Fault)

Proceed eastwards over the boulder strewn beach towards Yons Nab. One option is to follow the reef of Cornbrash limestone which runs through the bouldery area for some distance below Red Cliff. Alternatively, one can keep to the foot of the cliff, where a route tends to be tramped amidst the boulders. On reaching a small gully at the eastern end of Red Cliff, note that the Red Cliff Fault runs down the gully and that the Red Cliff Rock Member is now at the top of the cliff on the eastern (upthrown) side of the fault (Figs 58, 59). The throw here is about 37 m.

Yons Nab is composed of a gently, westerly dipping succession of Middle Jurassic strata lying beneath the Red Cliff Rock Member. The outermost reef is formed of the Millepore Bed, and it and the Yons Nab Beds run straight across the rock platform to meet the fault as it runs out northwards into the sea. All the overlying beds run from the rock platform into the cliff and rise gently as one proceeds east along the nab, and then descend back into the rock platform as one heads into Gristhorpe Bay. The succession is thus described here as it is seen during the traverse, starting at the top (Fig. 58).

7A. Western side of Yons Nab. The first beds one comes to in the low cliff section are Moor Grit Member sandstones and siltstones. There is a 5 m succession of massive beds of channel sandstone, each followed by delicately laminated silt and clay. The sandstones contain much carbonised wood. The Moor Grit again shows evidence of high sedimentation rates, with small-scale channeling followed by rapid infilling, and then pauses in sedimentation to deposit the laminated silt and clay. Ravenne (2002) has proposed that the sandstones at Yons Nab were situated on the south-eastern side of his 'Prism' II, being at the south-eastern margin of the palaeovalley infill which was centred in the area of the Peak Trough at South Bay, Scarborough (see below).

Below come 3.3 m of the Scarborough Formation. An upper unit consists of 2 m of soft, argillaceous, shelly limestones with a layer of concretionary ironstone (White Nab Ironstone) near the top. *Lopha* and *Meleagrinella* are abundant. Numerous small-scale faults disrupt the junction of the argillaceous limestone and the Moor Grit. Below is the Helwath Beck Member (Gowland & Riding, 1991), consisting of 1.3 m of delicately laminated siltstone with the bedding picked out by carbonaceous layers. Bioturbated layers, delicate ripple crests and swales, and scour-and-fill structures are beautifully displayed (Fig. 60).

The Scarborough Formation is very attenuated at Yons Nab compared with the sequence, probably 12 m thick in all, present at the centre of Cayton Bay. The Hundale Shale, Hundale Sandstone, Spindle Thorne Limestone and Ravenscar Shale members, most of which are present in Cayton Bay, are absent at Yons Nab. The Cayton Bay exposure lies within the Peak Trough (Milsom & Rawson, 1989), a fault-bounded trough operating in Mid Jurassic times which allowed thicker sequences of strata to accumulate within the trough than on its flanks. Particularly during deposition of the Scarborough Formation, subsidence on the eastern flank of the trough, east of the Red Cliff Fault, was much less than within the trough. In addition the Scalby Formation, Abbotsbury Cornbrash Formation, Hackness Rock Member and Yedmandale Member all show some attenuation in thickness in the area east of the Red Cliff Fault. An erosive junction of White Nab Member on Helwath Beck Member is visible in Gristhorpe Bay (see below). Ravenne (2002) proposed that the Peak Trough faults continued to be active during deposition of the Moor Grit, concentrating channel sandstones on their downthrown side in the area occupied by the trough.

The silts of the Helwath Beck Member rest on 1.5 m of silty, carbonaceous shale resting on a rootlet siltstone (0.7 m). Beneath the rootlet bed comes a series of 2.5 m of shales, grey above and sandy in part, but increasingly carbonaceous below, with at its base the well-known Gristhorpe Plant Bed (2.5 m). This consists of thinly laminated, friable clays

Figure 60. The Helwath Beck Member (Scarborough Formation) at Yons Nab. This marine horizon shows finely laminated scour structures and a thin bioturbated bed.

containing abundant plant debris and well preserved stems and leaves of *Bennetitales*, *Ginkgoales*, conifers, Ferns, Pteridophytes and *Caytoniales* (Fox-Strangways, 1892; Harris, 1961; Konijnenburg-van Cittert & Morgans, 1999). The plant bed is often largely obscured by cliff falls and beach gravel on the west side of the point, and is usually best seen on the east side in Gristhorpe Bay (Fox-Strangways, 1892).

Beneath the plant bed comes a sequence of 3.7 m of marginal-marine sandstones and silty, carbonaceous clays typifying the Gristhorpe Member. Three beds of sandstone are present, the highest rooted to a depth of 40 to 50 cm beneath the plant bed. Its base has infilled grooves and borings cut into the underlying clay. The middle sandstone is intensely contorted and injected into the underlying clay. The lowest one is markedly bioturbated. In between, the silty clays have flaser bedding, showing accumulation in an intertidal area. Below the lowest sandstone comes 1 m of grey, more typical alluvial shale containing much fossil wood including large, carbonised tree branches.

All the underlying beds at Yons Nab belong to the Lebberston Member, represented here by the Yons Nab Beds (Bate, 1959) deposited under somewhat marginal-marine conditions, underlain by the fully marine Millepore Bed. The highest part of the Yons Nab Beds consists of two beds of massive, iron-rich sandstone 90 cm and 120 cm thick, separated by 25 cm of shale. The upper sandstone is bioturbated, with pyrite concretions. The lower bed contains ironstone concretions which enclose marine bivalves, and only this bed was included in the Yons Nab Beds by Bate (1959).

7B. Horse Shoe Rocks. All the beds described above can be followed from the rock platform into the cliff. The remainder of the Yons Nab Beds run across the rock platform just east of the Nab. They consist of 5 m of flaggy alternations of shale and siltstone with numerous small, sideritic concretions and a marine bivalve fauna. The Millepore Bed forms the massive rampart at the seaward end of the rock platform. The highest 2 m consists of oolitic limestone with cross-bedding detectable beneath the barnacle-encrusted surface. Beneath, 7 m of cross-bedded sandstone is seen at very low tides, resting on fluvial sandstone of the Sycarham Member. The Millepore Bed varies noticeably in thickness across the nab (Livera & Leeder, 1981, fig. 4).

7C. Western end of Gristhorpe Bay. The whole sequence can now be followed back in stratigraphic order into Gristhorpe Bay. Of particular note is a Gristhorpe Member channel cutting out much of the Yons Nab Beds sequence east of the nab (Livera & Leeder, 1981, fig. 2). This sandstone has a slightly arcuate form of dipping beds deposited on a channel meander. The Gristhorpe Plant Bed is situated immediately above the channel sandstone (Fox-Strangways, 1892). In the cliff and rock platform there are good exposures of the Scarborough Formation displaying an abundant bivalve fauna. Two centimetre diameter burrows infilled with bioclastic sediment descend several centimetres from this fossiliferous bed into the Helwath Beck Member siltstone. A large channel in Moor Grit Member sandstone with persistent, westerly dipping foresets and strongly cross-cutting level-bedded siltstones, was first illustrated by Black (1928, fig. 1).

Locality 8. Gristhorpe Bay (Scalby, Cornbrash, Osgodby and Oxford Clay formations)

Proceed now to the centre of Gristhorpe Bay [TA 088 837]. A sequence from the Scalby Formation sandstones and siltstones to the Oxford Clay Formation can be examined in the cliff, but this does involve some scrambling. The chief interest of this exposure lies in the variety of fallen blocks of strata lying on the beach from which very varied faunas can be collected. Romano and Whyte (2003) illustrated a fallen block of Scalby Formation sandstone showing intense dinoturbation. The three thin limestone units of the Abbotsbury Cornbrash Formation can be distinguished: the basal, brown-weathering sideritic limestone, the middle pale grey, fine-grained micritic limestone containing *Trigonia* and many gastropods, and the upper bioclastic limestone containing bivalves and *Macrocephalites*. This is the best locality for Middle Callovian Langdale Member ammonites, the fallen blocks of fine- to medium-grained, bioturbated sandstone yielding *Erymnoceras*, *Perisphinctes* and *Kosmoceras grossouvrei* (Wright, 1968a). The Hackness Rock Member is a pale grey limestone with scattered chamosite ooliths and yields *Kosmoceras spinosum*, *Quenstedtoceras* spp. and *Collotia* sp. Blocks of Oxford Clay yield well-preserved if flattened *Cardioceras praecordatum* and *Parawedekindia arduennensis* (Wright, 1983).

Return to the NW end of Gristhorpe Bay. There used to be an easy access path up the cliffs here, but severe landslipping in 2016 carried the path away. It is possible to make one's way up the grassy cliffs on the eastern side of the nab to reach the Cleveland Way at the cliff top. It must be emphasized that there is no recognized path up here and some scrambling may be involved. This area has been showing signs of landslipping and it may be necessary to return to the north side of Yons Nab, where a rough path leads up to the Cleveland Way near Red Cliff. The route is then west over Red Cliff, offering splendid views of the coastal sections, and along the Cleveland Way either to the car park or to the access road to Cayton Bay Waterworks, which leads up to Osgodby Hill. The alternative is to return along the shore to Cayton Bay.

ITINERARY 10: BETTON FARM, EAST AYTON AND FILEY BRIGG

J.K. Wright and E.R. Connell

OS   1:25 000 OS Explorer 301 Scarborough, Bridlington & Flamborough Head
          1:50 000 Landranger 101 Scarborough
GS   1:50 000 Sheet 54 Scarborough

The itinerary is designed to demonstrate the relationships of sedimentary facies and the fossils contained in the varied series of limestones and marine sandstones which make up the Coralline Oolite Formation of the Corallian Group in NE Yorkshire. A cross section through the Corallian Group is given in Figure 61. The itinerary focuses on the Corallian Group sequences at Betton Farm, East Ayton and at Filey Brigg. It also describes the succession of glacial sediments well displayed in the wave-cut platform and cliffs on the south side of the Brigg. The Jurassic sections are described by J.K. Wright and the Quaternary by E.R. Connell.

Locality 1. Betton Farm Quarries (Malton Oolite Member with the Betton Farm Coral Bed)

The quarries are situated on the north and south sides of the busy A170 between Ayton and Scarborough, immediately to the east of East Ayton village. Together they form the Betton Farm Quarries SSSI. The southern quarry is owned by Basics Plus, a charity based at Betton Farm, now a tea room and farm shop. Parking is available here: permission to visit the southern quarry, adjacent to Betton Farm, must be obtained from Basics Plus by telephoning 01723 863143. A full description of the successions here was given by Wright and Rawson (2014).

The two quarries excellently expose a localised, coralliferous facies of the Malton Oolite Member. Wright and Rawson (2014) showed that the Betton Farm Coral Bed lies very near the base of the Malton Oolite Member and it therefore does not represent the much younger Coral Rag Member, as suggested by some earlier authors. In fact, Hudleston (1878) was the first to note that the characteristic Coral Rag echinoid *Cidaris florigemma* is not present here.

In Betton Farm South Quarry [TA 001 855], bioclastic sand, consisting largely of coral and bivalve debris, interdigitates with the development of a small ribbon reef made up of the massive colonial coral *Thamnasteria*. This ribbon reef, only 3 to 4 m wide and some 20 to 30 m long, was colonised by a fauna of reef-specializing bivalves and echinoids and sheltered a lagoonal area that accumulated lime mud and was colonized by large gastropods and burrowing bivalves (Fig. 62).

Approaching the South Quarry from Betton farm, the western face shows level-bedded coral-shell sand abutting the reef limestone of the ribbon reef (Fig. 63F). At the junction of the two, the initial growth of a large mound of *Thamnasteria* (Fig. 63D) was followed by a period of encroachment of coral-shell sand over the massive coral, with subsequent growth of coral over the coral-shell sand.

Proceeding to the centre of the quarry, on the east side of the 3 to 4 m thick ribbon reef, one is entering the sheltered lagoon (Fig. 62). Close to the ribbon reef, coral-shell sand washed in and organisms which lived in close proximity to coral thrived. This variant of the protected facies consists of coarsely shelly micrite with unbroken *Chlamys nattheimensis*, *Lopha gregaria* (Fig. 63C) and *Nanogyra nana*, spines of *Paracidaris*

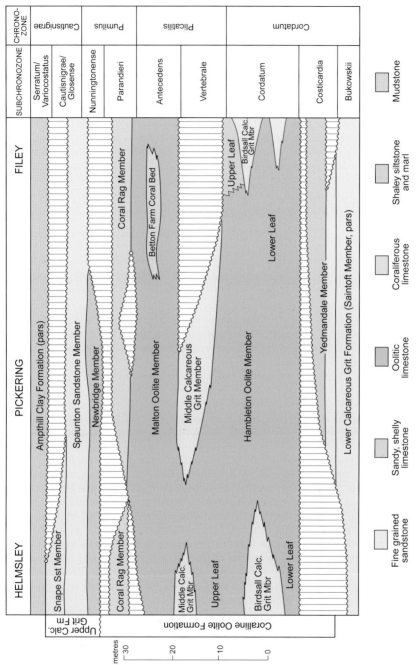

Figure 61. Schematic cross section of the Corallian rocks of the Vale of Pickering. Modified from Rawson and Wright (1996, fig. 15).

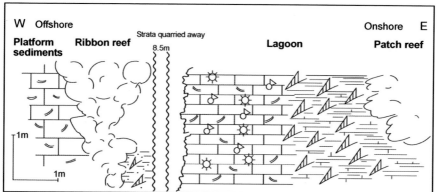

Figure 62. Reconstructed cross section of Betton Farm South Quarry. From Rawson and Wright (2014, fig. 13), reproduced by kind permission of the Yorkshire Geological Society.

(*Anisocidaris*) *smithii*, *Rhabdophyllia phillipsi* and numerous small, abraded fragments of *Thamnasteria* and *Thecosmilia*.

On the eastern side of the quarry, these shelly carbonates pass laterally into the argillaceous, micritic, marly limestone which accumulated in this very protected area. A substantial population of the gastropod *Bourgetia saemanni* became established (Fig. 63A). There are scattered thin-shelled bivalves (*Modiolus* sp.) and echinoids such as *Pseudodiadema* sp. and *Hemicidaris* cf. *intermedia*. Patches of *Thamnasteria concinna* then expanded laterally from both east and west sides into the lime mud area.

Betton Farm North Quarry [TA 002 856] exposes the sediments laid down prior to colonization of the area by corals, these sediments being overlain by separate, discrete colonies of *Thamnasteria*. At the base of the quarry is a bed of brown-weathering, argillaceous oolite with numerous *Nanogyra nana*. Only rolled fragments of *Thamnasteria* and *Thecosmilia* are found in the overlying shelly oomicrite, but this is followed by a 20 cm thick layer of spary oobiomicrite with rolled, but not abraded, fragments of the delicate branching coral *Rhabdophyllia phillipsi*. Separate, discrete colonies of *Thamnasteria* then became established. These isolated, dome-shaped masses surrounded and extended over the coral-shell sand. Various authors have recorded a substantial fauna from these inter-reef beds, comprising at least six species of gastropods, three of echinoids, three of brachiopods and five of bivalves. The coral mounds thus grew in the back-reef sand belt, rising above the bare lime-sand foundation in moderately high-energy conditions, the area sufficiently sheltered by a framework reef situated to the west to allow a substantial epifauna to become established.

Locality 2. Filey Brigg (uppermost Lower Calcareous Grit Formation, lower part of the Coralline Oolite Formation and Devensian tills)

The Saintoft Member at the top of the Lower Calcareous Grit Formation, the Yedmandale, Hambleton Oolite and Birdsall Calcareous Grit members of the Coralline Oolite Formation, and the overlying succession of glacial sediments, are well displayed in the wave-cut platform and low cliff of the Brigg. The map (Fig. 64) shows the geology of the area and the recommended stopping places. Parking is available in the Filey Brigg Country Park (pay) car park [TA 119 811]: park near the shop and walk down the track opposite, past the sailing club to the shore.

Figure 63. (facing page) Betton Farm South Quarry and its fauna. (A) *Bourgetia saemanni*. (B) *Paracidaris* (*Anisocidaris*) *smithii* spines. (C) *Lopha gregaria*. (D) *Thamnasteria concinna*. (E) Close-up view of part of the face shown in Fig. F, showing encroachment of coral-shell sand (left) over massive *Thamnasteria* (middle of hammer shaft) and subsequent growth of the coral over the sand (above hammer). (F) The western face of the quarry, showing level-bedded coral-shell sand facies abutting against reef limestone to the right. Figures 62D–F are reproduced from Wright and Rawson (2014, figs. 9, 12 and 11) by kind permission of the Yorkshire Geological Society.

1. The Jurassic sequence at Filey Brigg

The log (Fig. 65) shows the complete succession, though no more than 3–4 m of this are accessible at any one point as the beds dip gently in the low cliffs. The stratigraphic terminology is that of Wright (1983). Wilson (1949) published a very detailed measured section of the Coralline Oolite Formation at Filey Brigg and this is very useful to the enthusiastic collector, although the stratigraphic terminology is very outdated. Wilson divided the succession into 34 beds and extensive fossil lists were given for each bed. It is however difficult to match Wilson's beds precisely with those seen at the Brigg, though an attempt was made recently by Coe (1996, fig. 14). The numbers used in the log here are those of Wright (1983).

The log brings out the predominantly sandy nature of the succession at Filey Brigg. There are just three beds of standard oolitic limestone, all in the Lower Leaf of the Hambleton Oolite Member. The principle lithology is a calcareous sandstone, particularly those of the Birdsall Calcareous Grit Member (nos. 7 to 11). This sandstone, formerly referred to the Middle Calcareous Grit, contains an ammonite fauna identical to that found in the type quarry sections at Birdsall, near Malton. The Birdsall Member comprises a wedge of sand poured into the southern side of the Cleveland Basin during an uplift of the Market Weighton High. Oolite (Hambleton Oolite Member) continued to be deposited throughout in the northern half of the basin, but where the Birdsall Member is present the Hambleton Oolite is divided into Lower and Upper Leaves (Fig. 61).

1A. North side of the Brigg (Fig. 66). On reaching the shore, turn left across the sands and head towards the Brigg, along a concrete path and over rocky scars. The going is fairly straightforward, though the algae-covered parts of the path and scars can be very slippery. Continue to the vicinity of a small hut (bird hide) then scramble over boulders onto the slightly lower rock platform and walk to the small bay about 100 m NW of the hide. From here the succession can be followed upwards on the platform and adjacent cliff back to the hide. The rock platform in the embayment displays the round 'cannonball' concretions so typical of the Saintoft Member of the Lower Calcareous Grit Formation.

Resting on an erosion surface cut in the Saintoft Member, the lowest 0.6 m of the Yedmandale Member consist of a fine- to medium-grained sandstone. The sandstone is heavily bioturbated and contains *Nanogyra nana* and *Chlamys fibrosus*. Above comes the main Yedmandale Member limestone, 2 m of grey-weathering limestone in 5 or sometimes 6 beds (for faunal list see Wilson (1949), beds 1–8). *Nanogyra* colonies weather out, and there are many dissociated *Gervillia valves* in the upper beds and some in life position in the top bed. Small-scale cross-bedding fills small scours and dips to the south. These Yedmandale Member limestones seem to have been deposited in a series of storm surges washing coral and shell debris from shallow water to the north-west into the offshore shelf area to the south-east (Wright, 1992, fig. 10). There were thus brief episodes of sedimentation in an area normally accumulating little sediment. Only in the highest bed is the fauna indigenous. Compared with the 2.6 m here, the Yedmandale Member is at least 10 m thick at the type locality (Wright, 1992).

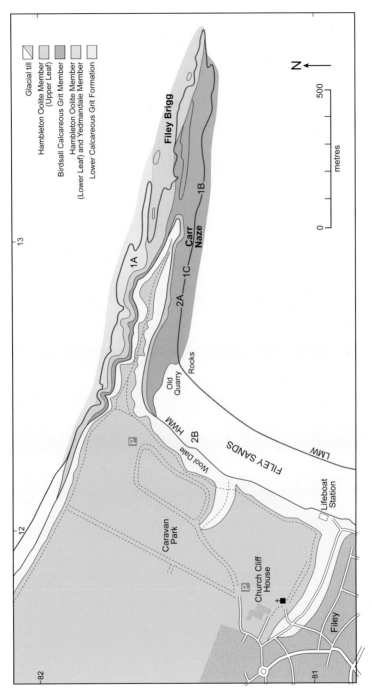

Figure 64. Map of the Pleistocene and underlying Jurassic strata at Filey Brigg (Itinerary 10).

3 Excursions: Itinerary 10

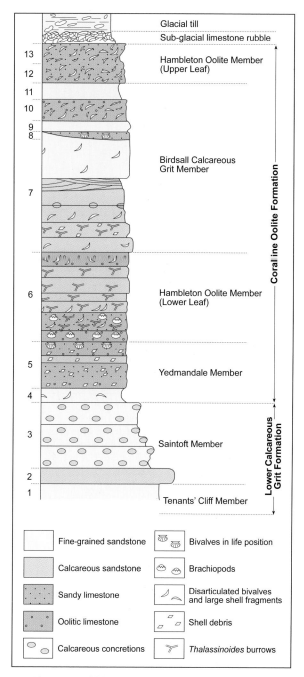

Figure 65. Lithic log of the Corallian at Filey Brigg.

113

Figure 66. The Corallian sequence at Filey Brigg (Yedmandale to Hambleton Oolite members), overlain by the lower till unit.

The next 3.7 m of limestone belongs to the Lower Leaf of the Hambleton Oolite Member. A major bedding plane marks the base, above which comes the first massive bed of oolite. None of the oolites at Filey show cross-bedding and they seem to have accumulated in a fairly stable, quiet environment allowing the excellent preservation of delicate echinoids and brachiopods (for faunal list see Wilson (1949) beds 9–14 only). The quiet conditions also favoured the development of extensive networks of *Thalassinoides*. The infilled burrow systems weather out in spectacular fashion in the large, fallen blocks towards the centre of the Brigg (Fig. 67). The higher, sandy limestones of the Lower Leaf form the NW side of the Brigg.

1B. Filey Brigg. The Brigg itself is formed of the tough, calcareous sandstones of the Birdsall Calcareous Grit Member. The full thickness is 6.8 m. Towards the base there are calcareous concretions with shelly bands containing occasional *Cardioceras* spp. Massive, occasionally cross-bedded, sandstone forms the bulk of the unit with, at the top, two beds of tough, calcareous sandstone. Fossils from these beds can be collected loose from the gravel and boulders on the south side of the Brigg.

1C. South side of Filey Brigg. The higest beds of the Birdsall Calcareous Grit (8–11) form the rock platform and lower part of the cliff here. In the corner of the bay at Old Quay Rocks [TA 816 125], *Cardioceras persecans*, a typical representative of the Lower Oxfordian Cordatum Zone, can be collected from the calcareous sandstone of Bed 8.

3 Excursions: Itinerary 10

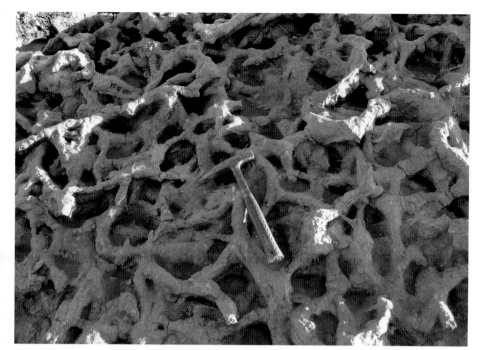

Figure 67. *Thalassinoides* systems etched out in a loose block of the Lower Leaf of the Hambleton Oolite Member, Filey Brigg.

Continuing up the succession, there is a marked change in lithology from bed 9 (yellow, massive sandstone) into Bed 10 (tough, shelly limestone with many serpulae). However, there is then a return to sandy facies (Bed 11), a laminated, shelly sandstone. This contains numerous infilled *Thalassinoides* descending from the shelly limestone (Bed 12) which is now, following Coe (1996), taken as the base of the Upper Leaf.

The Upper Leaf of the Hambleton Oolite is excellently seen in a continuous section along the south side of the Brigg. It consists of tough, impure very fossiliferous, bioclastic limestone containing well-preserved bivalves and occasional *Cardioceras excavatum*, *C. (Subvertebriceras)* spp. and *Perisphinctes* sp. The appearance of *Perisphinctes*, a genus which occurs only very rarely in the underlying strata, along with *C. excavatum*, is an indication that the highest Hambleton Oolite exposed at the Brigg is of Middle Oxfordian age, belonging to the Vertebrale Subzone. However, only 1.5 m of this Upper Leaf is seen beneath the limestone rubble at the base of the thick tills that form the bulk of the cliff here.

2. Glacial deposits
While several North Yorkshire coastal embayments contain significant thicknesses of glacial sediments deposited during the last glaciation (the Dimlington Stadial) most are generally poorly exposed at present (see Catt, 2007; Catt & Madgett, 1981). However, at Filey Brigg the tills are well exposed, thick (30–40 m) and display a range of glacigenic features. The sections have been described and interpreted by a number of workers (Edwards, 1981; Evans *et al.*, 1995; Boston *et al.*, 2010).

2A. Base of the till sequence at Carr Naze. On the south side of the Brigg, the base of the glacial sequence can be seen at TA 130 815. Here, Jurassic limestone has been brecciated and glacitectonically folded, with a sense of movement from east to west, i.e. apparent stress from the east, mirroring the fabrics of clasts in the overlying tills (Evans *et al.*, 1995) and indicating ice advance varying from the E or NNE. In early work on the glaciation of the site both Lamplugh (1891) and Stather (1897) noted striae engraved into the Jurassic bedrock indicating ice advance from the NNE.

2B. Old Quay Rocks and Wool Dale. The glacial sequence can be seen most conveniently from the beach close to Old Quay Rocks [TA 125 816] on the south side of the Brigg (Figs 68, 69). Here, above the Jurassic rocks, the lower till, which is grayish-brown in colour, makes up the lower half of the cliff. Overlying the till are a number of 'ledges' composed of gravels and sands extending significant distances along the cliff exposures. Uncertainty surrounds the deposition of these units (see discussion in Boston *et al.*, 2010). They have been interpreted as outwash sediments deposited as the ice sheet withdrew but prior to a readvance depositing the upper till unit. Alternatively, they may have been deposited in subglacial channels implying no ice-free period as the glacial sediments aggraded at the site. Above the gravels and sands the upper part of the cliffs is composed of a further till unit with a distinctive reddish-brown matrix colour suggestive of a change in provenance of the entrained sediments. The lower and upper till units at Filey Brigg have

Figure 68. View east along the cliffs on the south side of Filey Brigg displaying the stratigraphy of the glacial sediments near Old Quay Rocks. Lower and upper till units are separated by a number of gravel and sand units standing out as ledges in the section.

Figure 69. Glacial diamicton facies in the lower till unit at the cliff base at Wool Dale (just to the west of Old Quay Rocks). Note the deformation structure and laminated/bedded diamicton facies at the base of the section passing up into more massive diamicton with sporadic large erratic clasts.

been most recently correlated with the Withernsea and Skipsea tills respectively of the Holderness coastal sections (for more details see Holderness itineraries below).

Whilst both till units seem massive in stucture from a distance, and are generally rather inaccessible in the steep cliffs, it is possible to approach the lower, grayish-brown till unit at the base of the cliff just to the west of Old Quay Rocks in the area marked as Wool Dale on the OS 1:25 000 scale map [TA 124 815] (Fig. 69). At this site the till is composed of both massive and laminated/bedded diamictons which are locally deformed and clearly folded towards the base. Wave action appears to have eroded out some of the thin beds of sand/mud highlighting these structures. As the ice sheet advanced into the area it appears to have deformed fine-grained sediments (proglacial lacustine facies?) and glacitectonically incorporated them into the basal zone of the subglacial till. It is unclear if similar structures are present in the other till facies where they have not been exposed to wave action.

ITINERARY 11: REIGHTON GAP TO SPEETON CLIFFS

P.F. Rawson

OS   1:25 000 Explorer 301 Scarborough, Bridlington & Flamborough Head.
         1:50 000 Landranger 101 Scarborough
GS   1:50 000 Sheet 55/65 Flamborough and Bridlington

This itinerary is a full-day excursion primarily to study the type section of the Lower Cretaceous Speeton Clay Formation and overlying chalks. The maximum walking distance is about 5 km, mainly over sand and shingle, though the shore becomes boulder strewn near to the chalk cliffs. To examine the whole sequence, turn off the coast road (A165) onto the road to Reighton village and just north of the village [at TA 128 757] go onto a minor road signposted for Reighton Sands holiday village (Fig. 70). Follow this road until it forks in front of the holiday camp, then take the left branch to a small parking area [TA 140 763]. From here take the signposted public footpath to the beach and turn right, walking round a small point, towards the chalk cliffs in the far distance. There is a second access to the shore down a private road from the holiday camp.

While the shore from here to Middle Cliff is generally sandy, occasionally patches are stripped off to expose either till or disturbed Kimmeridge Clay. The adjacent cliffs are of brown and reddish coloured tills (Skipsea Till) and show numerous landslips. The tills are the source of the numerous erratic beach pebbles.

Locality 1. New Closes Cliff (Kimmeridge Clay and Speeton Clay formations)

About three quarters of a kilometre from the footpath, opposite the south-eastern end of New Closes Cliff, is a cluster of concrete blocks: on the rare occasion when sand has been removed from this area a small synclinal relic of the lower part of the Speeton Clay D beds is exposed (Fig. 70). Just beyond, in the vicinity of a concrete breakwater that effectively marks the boundary between New Closes and Middle cliffs (Fig. 70, BW1), the topmost paper shales of the Kimmeridge Clay Formation are sometimes visible at the cliff foot or on the adjacent shore. Rarely, a band of large septarian concretions with attractive greenish-yellow calcite crystals is exposed. Flattened ammonites and small bivalves (*Lucina minuscula*) occur in both the concretions and the shales, the ammonites representing the Hudlestoni to lower Pectinatus zones (Table 3). The shales are tightly folded into small, E-W trending, very angular anticlines and synclines, reflecting a Cenozoic compressional phase.

Locality 2. Middle Cliff to Speeton Beck (Speeton Clay Formation)

2A Middle Cliff. Here, dark grey clays appear from beneath the tills in the cliff face, marking the commencement of the outcrop of the Lower Cretaceous Speeton Clay Formation, which rests disconformably on the Kimmeridge Clay Formation. The Speeton Clay Formation, only about 109 m thick here, is exposed in Middle and Black cliffs for about 0.75 kilometre south-eastward to Speeton Beck. The cliffs are unstable and continuously changing, so that rarely is the whole extent cleanly washed by the sea; instead there are often slipped clays along the cliff foot which can be very treacherous in wet weather. Conversely, patches of shingle and sand are sometimes stripped off the intertidal zone to give very clear exposures of the clays beneath. Again, the clays show small, sometimes tight, E-W folds.

3 Excursions: Itinerary 11

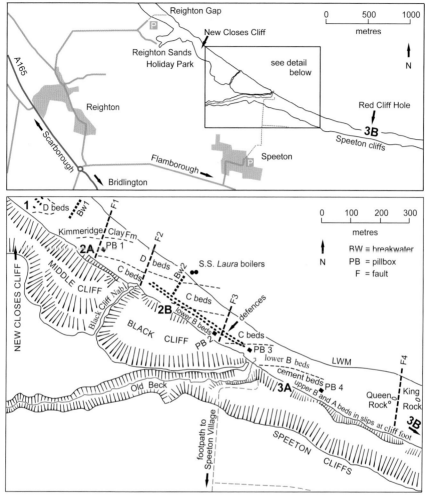

Figure 70. Map of the Speeton section (Itinerary 11).

The Speeton Clay Formation was divided by Lamplugh (1889) into 4 units, the A–D beds (labelled from the top downward: Table 4), characterised by abundant belemnites of alternating northern (Boreal (B)) and southern (Tethyan (T)) origin:

A beds (c. 12 m) - *Neohibolites* (T)
B beds (c. 46 m) - *Praeoxyteuthis*, *Aulacoteuthis* and *Oxyteuthis* (in ascending order) (B)
C beds (39 m) - *Hibolites* (T)
D beds (12 m) - *Acroteuthis* (B)

Finer lithological units were distinguished by Lamplugh and subsequent workers. Only the D, C and lower B beds are well exposed, and a simplified lithic log for these is given in Fig. 71. Although it is difficult to follow the section in detail unless it is very cleanly

3 Excursions: Itinerary 11

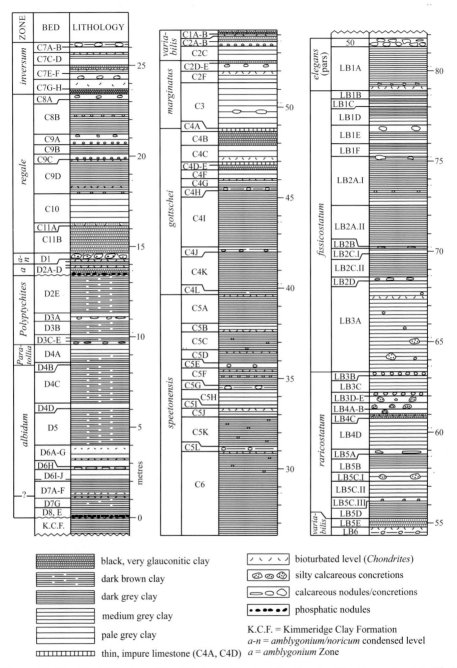

Figure 71. Simplified lithic log of the D to lower B beds of the Speeton Clay Formation. Modified from data in Neale (1960, 1962), Kaye (1964), Fletcher (1969) and Rawson and Mutterlose (1983).

exposed, there are several distinctive marker horizons which one can look for as a key to reading the sequence. The first is at the base of the clays, where a 10 cm thick phosphate nodule bed (bed E, the Coprolite Bed) with derived Kimmeridge Clay fossils marks re-working and a break in deposition over some 9 million years. This bed was mined as a source of phosphate until a major landslip closed the mines in 1869 (Lamplugh, 1889). Pit props are still uncovered occasionally at the cliff foot and help to locate the bed; where it has been mined away the immediately overlying clays are disturbed by slumping.

The commencement of deposition of the Speeton Clay Formation is the local reflection of an important sea-level rise that flushed out the whole North Sea Basin (Rawson & Riley, 1982). From then on sedimentation continued slowly, with occasional interruption, through the remainder of Early Cretaceous time.

If the Coprolite Bed is not visible the first obvious marker is usually the pale but bright, striped clay of D6. The overlying bed D5 is a brackish-water deposit which contains the primitive brachiopod *Lingula*, often preserved in its pyrite-infilled burrow. D4 marks a return to fully marine conditions and contains numerous bivalves, mainly *Astarte senecta* and the massive oyster *Exogyra latissima*. A brown-weathering silty clay with irregular concretions marks the top of D3 (D3A) and sometimes yields partially crushed large ammonites belonging to the Boreal genus *Polyptychites*. A very distinctive band of large calcareous concretions enclosing smaller phosphatic nodules, bed D1, the 'compound nodule bed' (Fig. 72), is another clear marker. It reaches the shore close to a large slab of concrete (sometimes covered by sand) that once formed the base of a Second World War pillbox (Fig. 70, PB1). About a metre below D1 is a phosphatic nodule horizon (base of D2D) marking an important break in the sequence; the whole of the Upper Valanginian substage is cut out here, Lower Hauterivian clays resting on Lower Valanginian ones. Phosphatised Upper Valanginian ammonites occur among the nodules, together with corroded and water-worn *Acroteuthis*.

Figure 72. Beach exposure of bed D1, the 'compound nodule bed', Speeton. Small brown, partially phosphatised nodules are enclosed in hard, bioturbated calcareous concretions.

Bed D1 itself is another condensed horizon rich in fossils, including *Acroteuthis*, ammonites (*Endemoceras* and *Distoloceras* up to 0.7 m in diameter) and the 'shrimp' *Meyeria ornata*. In the overlying C beds, the Tethyan-derived belemnite *Hibolites jaculoides* is common throughout, while ammonites of both Boreal and Tethyan origin occur. The latter enable correlation of some levels with the 'standard' sequences of southern Europe (Kemper *et al.*, 1981; Rawson, 1995, 2014) and further afield (Rawson, 2007). Beds C11–C8 are quite fossiliferous, with attractively preserved small ammonites (including *Endemoceras regale*, *Olcostephanus* and *Parastieria peltoceroides*), *Hibolites jaculoides* and *Meyeria ornata*. A new ammonite fauna appears in C7, where the Boreal genus *Simbirskites* (subgenus *Speetoniceras)* is common at the base (Rawson, 1971) and the uncoiling (heteromorph) ammonite *Aegocrioceras* (Rawson, 1975) just above: the latter is endemic to north-west Europe but derived from a Tethyan genus, *Crioceratites*.

About 50 metres along the cliff from the D1 outcrop is a thin, bright yellow sulphurous band about 3 m above the shore that forms a very clear marker zigzagging along the remainder of Middle Cliff (Fig. 73). This has weathered from a very thin pyrite-rich layer of probable volcanic origin within bed C7E and is not visible when the clay is freshly exposed on the shore. Just beneath it is bed C7F, the first of several silty, reddish-weathering bands that form distinctive markers higher up the succession. The first three (C7F, C7A and C5L) can be reached via gullies in the cliff. The higher ones (C5G, C4C, C4J

Figure 73. Three distinctive marker horizons in the mid C beds, Speeton. The yellow-weathering (sulphur) streak is a very thin horizon, probably weathered volcanic ash, in bed C7E: just beneath is the brown-weathering, silty bed C7F, while a similar bed about 1m above the yellow streak is bed C7A.

and C4H) are best seen on the rare occasions when the nearby beach is stripped off. Bed C7F contains common body-chambers, sometimes with inner whorls, of *Aegocrioceras quadratum*, while C7A yields Tethyan *Crioceratites*. The clays of C6 contain small *Simbirskites*, especially near the base and top.

2B. Black Cliff. Middle Cliff terminates at a ridge (Black Cliff Ridge) that coincides closely with a fault (F2) crossing cliff and shore just before the start of a line of concrete blocks paralleling the cliffs. To the SE of the fault a low vertical cliff exposes the upper C and lowest B beds, a series of thinly bedded, cyclic units in which the 10 cm thick, intensely glauconitic bed C2D is a good marker, as are the bioturbated clays of C1 and another intensely glauconitic clay forming bed LB5E (0.46 m). Fossils are not common but the section is interesting for the occurrence of the small brachiopod '*Terebratulina' martiniana* in the glauconitic clays of bed C2A (see Middlemiss, 1976, p. 75).

Starting only a few metres further along Black Cliff, the immediately succeeding lower B beds section described by Rawson and Mutterlose (1983) is patchily visible though partly obscured by slipped clay. The most distinctive marker beds include the thin (0.23 m) very glauconitic black clay of LB4C, the calcareous silty doggers of LB3E to LB4B and the two beds of small round calcareous nodules of beds LB3D and LB3B, which are separated by about a metre of clay. These clays are not very fossiliferous, though the belemnites *Praeoxyteuthis*, *Aulacoteuthis* and *Oxyteuthis* occur in upward succession. Rare heteromorph ammonites occur, either as flattened calcareous films or more solitary fragments.

The junction between lower B and the cement beds is exposed nearly opposite a second breakwater of concrete blocks (BW2). Here, very dark, laminated, kerogen-rich shaly clays are visible just beneath a distinctive double cementstone band that is sometimes visible on the shore at the cliff foot and marks the base of the cement beds. These clays (LB1A) are very pyritic and attractive small crystals occur in clusters. Pyritised whorl fragments of the heteromorph ammonite *Paracrioceras elegans* also occur (Rawson & Mutterlose, 1983), together with common examples of the gastropod-like serpulid tube *Rotularia*. Bed LB1A marks the top of a sequence of laminated clays (LB1A to LB1F) that can be traced across the North Sea to North Germany, where they are called the 'Blatterton' (= paper shale). They represent a brief anoxic event in Early Barremian times.

Where breakwater 2 runs out to sea two ship's boilers form a distinctive landmark. They form part of the wreck of the 2089 ton Austrian steamship *Laura*, which was sailing from Newcastle to Trieste with a cargo of coke when she ran ashore in dense fog on 21 November 1897 and broke in two (Godfrey & Lassey, 1974, p.73, photos, pp. 114–115). The wreck is sometimes exhumed from the sand, when both halves of the ship can be seen at very low tides.

Beyond the basal cement beds there is an extensive slipped area of till and red chalks. The Speeton Clay reappears 30 m SE of the remains of another, tilted pillbox base (PB2), where the top C beds and lower B beds sequence appears again, repeated by fault F3. The succession is exposed along the remainder of Black Cliff as far as Speeton Beck, though again often partially obscured by slipped clay.

Locality 3. Speeton Beck to Red Hole (Speeton, Hunstanton and Ferriby Chalk formations)

3A. Speeton Cliffs. Immediately to the south-east of Speeton Beck the Cement Beds crop out at the foot of the cliff though again they are often obscured by landslip. From here one can either return along the shore to Reighton Gap, or continue south-eastward to examine the chalks of the Hunstanton and Ferriby formations beneath Speeton Cliffs. This second part of the itinerary can also be treated as a separate visit, in which case an alternative route is to park next to the church at Speeton village [TA 152 746] and follow the clearly signposted footpath from the church to the cliff top, then down the undercliff to the south-eastern side of Speeton Beck: the last part of this path is steep and can be slippery in wet weather.

To the SE of the Speeton Beck section, the Speeton Clay Formation is buried beneath landslipped chalk and till and has rarely been seen on the shore. Isolated small patches of the upper cement beds and upper B are sometimes uncovered at the cliff foot, all extensively disturbed by faulting or squeezing. Black pyritic clays (top upper B) are sometimes well exposed opposite Queen Rock, yielding Late Barremian and Early Aptian ammonites and the bivalve *Grammatodon securis*. Slickensided calcite, a bed of tectonically deformed belemnites and bedding-plane slips make the detailed sequence difficult to establish.

Further along, the higher shore is strewn with chalk boulders which make the walking more difficult; between them, patches of red and grey calcareous clay (A beds) are sometimes uncovered. Eventually the landslip area gives way to sheer chalk cliffs about a kilometre from Speeton Beck. Both in the cliffs and intermittently on the adjacent foreshore, the Hunstanton Formation (formerly 'Red Chalk') is exposed. The impure chalks and thin marls here are considerably thicker than the typical condensed limestone facies inland on the East Midlands Shelf and mark a gradation towards the equivalent but generally more argillaceous Rødby Formation of the offshore area. Fossils are moderately common, especially brachiopods and species of the small belemnite *Neohibolites*.

Mitchell (1995) divided the Hunstanton Formation at Speeton into 5 local members. His paper includes a location map and detailed lithic logs. The lower two members are poorly exposed, but the overlying Dulcey Dock and Weather Castle members are visible in the first cliff exposure, a low (possibly downfaulted) cliff face, and on the nearby foreshore. The Albian/Cenomanian boundary lies approximately 0.5 m below the top of the latter member (Mortimore *et al*., 2001, p. 418).

3B. Red Cliff Hole [TA 165 751] is a well-marked recess at the beginning of the high chalk cliffs. Here the upper part of the Weather Castle Member is exposed, overlain by up to about 3.7 m of greenish-grey chalk (the 'Grey Band'). The latter bed yields the Cenomanian brachiopod *Concinnithyris subundata*, plus large pyrite crystals and marcasite. It is overlain in turn by another reddish-coloured unit, some 2–3 m thick. Note that the change in colour is sharp but cuts across the bedding irregularly—the grey colour is apparently secondary, due to reduction of iron minerals. Both the 'Grey Band' and the overlying reddish beds were formerly included in the Chalk Group (forming the base of the Ferriby Chalk Formation), but Mitchell (1995) assigned them to the top of the Hunstanton Formation, as a fifth (Red Cliff Hole) member.

Above the Red Cliff Hole Member, the Ferriby Chalk Formation (*c*. 33 m) is again considerably thicker than in inland exposures. A detailed section is given by Mortimore *et al*. (2001, fig. 5.22). The lowest part of the sequence at Red Hole consists of about 2.2 m of white, flaser-bedded chalks and thin red-purple marls placed in the Crowe's Shoot Member (Mitchell, 1995). Most of the overlying sequence is only accessible beyond Red Hole, but higher in the cliff here two thin pink chalks can be seen approximately 6.5 and 25 m above the base of the Ferriby Chalk, while at 33 m the 'Black Band' (OAE level) at the base of the overlying Welton Chalk Formation is visible (Fig. 74).

3 Excursions: Itinerary 11

Figure 74. Red Hole, Speeton. The red chalks at the foot of the cliff form the upper part of the Weather Castle Member, while the Black Band is visible just beneath the grass-covered slopes.

The visitor should not walk beyond Red Hole as the sections are dangerous and the tide reaches the cliff foot in places. It is equally dangerous to scramble up the grassy slopes to reach higher beds.

ITINERARY 12: THORNWICK BAY AND NORTH LANDING, FLAMBOROUGH

P.F. Rawson

OS   1:25 000 Explorer 301 Scarborough, Bridlington & Flamborough Head
         1:50 000 Landranger 101 Scarborough
GS   1:50 000 Sheet 55/65 Flamborough and Bridlington

The localities described in itineraries 12 to 14 form part of the Flamborough Head SSSI and lie in the area of the Flamborough Headland Heritage Coast. When approaching sections visitors are asked to keep to the marked paths.

In the Flamborough area the coastline is deeply eroded from Thornwick Bay to Flamborough Head and there are several small bays which contain magnificent arches, caves and sea stacks cut into hard chalk—so hard that it has been used as a building stone and can be seen in some of the older buildings in the area, including the 17th century lighthouse. At low tide some of the caves and arches provide access to adjacent coves, from which there is no escape when the tide turns. **It is highly dangerous to stray beyond the confines of the bays described here.**

Throughout the Flamborough area the Chalk is overlain by a thick blanket of boulder clay (Skipsea Till) and it is the contrast in hardness between the two that causes such a marked change in slope half to three quarters of the way up the cliffs. In many places downwash from the clay smears the chalk while fallen lumps are soon broken up by the sea to release the enclosed erratic rocks and fossils. Although the beach shingle is predominantly of local flint and chalk it contains a variety of exotic pebbles. Erratic boulders are also scattered over the beaches—including once again the distinctive Shap Granite.

Locality 1. Thornwick Bay (Welton and Burnham Chalk formations)

From Flamborough village take the B1255 towards North Landing (Fig. 75), turning immediately before 'The Viking Hotel' onto a track signposted for the 'Thornwick Café': a small parking fee is sometimes payable at the entrance. About 400 m along the track is a small grass-covered, rough parking area opposite a narrow footpath leading down to Great Thornwick Bay. During the holiday season it is possible to drive along the track for another 300 m until it terminates at a parking area opposite the Thornwick Café, but when the cafe is closed this part of the track is too. Alternatively, there is extensive parking throughout the year at nearby North Landing (see Loc. 2 below).

Adjacent to the cafe is a grassy area providing excellent viewpoints over Great and Little Thornwick bays. In both bays the chalk is very sparsely fossiliferous and belongs almost wholly to the *Terebratulina lata* Zone of the Turonian. The exposures are of interest primarily to show how this single, ill-defined Chalk biozone can be finely subdivided lithologically and they provide an excellent exercise in 'proving the section' from a lithic log. The sequence embraces the upper half of the Welton Chalk Formation and the base of the overlying Burnham Chalk Formation and contains a number of named marl and flint bands (Fig. 76) that occur across much of the northern province. The sediments accumulated along the southern margin of the Cleveland Basin, within the Howardian-Flamborough Fault Belt, and are almost 10 m thicker than in correlative sections on the northern part of the East Midlands Shelf.

3 Excursions: Itinerary 12

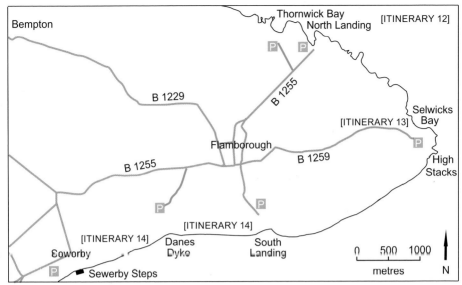

Figure 75. Locality map of the Flamborough area (Itineraries 12–14).

1A. Little Thornwick Bay. Avoid the footpath leading down from the viewpoint towards Thornwick Nab as it is narrow, slippery and dangerous and there is no access into Little Thornwick Bay. Instead, walk down a steep, partially stepped footpath on the far (west) side of the café and turn down into the bay. Here, the lowest beds seen in this itinerary are visible on the north side of the bay (Fig. 77), where four deeply eroded, narrow clefts near low-water mark (the lowest at the foot of the arch on the seaward side) represent the individual marl bands of the Barton Marls. Two metres above the highest marl is a 5 cm thick tabular flint band and 1 m above that is the Ferruginous Flint. This very distinctive bed is a prominent, up to 15 cm thick, tabular, carious flint with reddish-brown weathering patches. It can be traced on both sides of the bay and round Thornwick Nab into Great Thornwick Bay, forming one of the best marker bands. A second good marker is a 2–4 cm thick marl, the Melton Ross Marl, which forms a prominent groove in the cliff, reaching almost to shore level at the head of Little Thornwick Bay.

1B. Great Thornwick Bay. From Little Thornwick return towards the café and then walk back along the track for about 250 m to join a path leading down into Great Thornwick Bay [TA 234 720]. Cross the chalk scars towards Thornwick Nab with its distinctive arch (this should only be done on a falling tide). The roof of the arch is formed by the Ferruginous Flint, while beneath it the Barton Marls are exposed in the lower part of the nab (Fig. 78).

The Ferruginous Flint and the thinner tabular grey flint 1 m below can be traced in the adjacent cliff dipping gently down towards shore level to cross the intertidal rock scars in the northern part of the bay, about 25 m from the nab. From here the succession can be traced upward across the scars to the southern and eastern sides of the bay (Fig. 79). The next flint above the Ferruginous Flint is a nodular flint, some of the nodules showing the distinctive shape of *Thalassinoides*. About 3 m higher in the sequence is a bed of larger flint nodules, some again infilling *Thalassinoides*. In places the nodules coalesce into

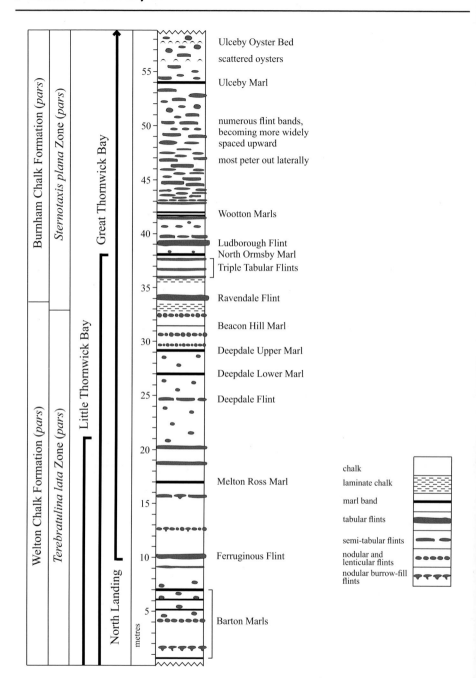

Figure 76. Lithic log of the Chalk at Thornwick Bay and North Landing. Logged by P.F. Rawson and the late F. Whitham.

3 Excursions: Itinerary 12

Figure 77. The Chalk at Little Thornwick Bay. The Barton Marls (BM) occur at the foot of the cliff: the Ferruginous Flint (FF) is 3 metres above, while the Melton Ross Marl (MR) forms a notch higher in the cliff.

Figure 78. Thornwick Nab. The Ferruginous Flint (FF) is the prominent band half-way up the cliff, forming the top of the arch.

3 Excursions: Itinerary 12

Figure 79. View across Great Thornwick Bay, showing the higher part of the Welton Chalk and the lower part of the Burnham Chalk formations.

sheets of flint up to a metre across. The overlying Melton Ross Marl forms a slight notch running at the base of a scar some 15 m from the adjacent cliff face on the south-western side of the bay. While the flint bands normally stand out from the enclosing chalk, the opposite is the case with the two thin bands above the Melton Ross Marl. They form the lowest part of the vertical face of two scars close to the cliff and are easily overlooked.

In this south-western cliff, the lowest flint band is a prominent grey semi-tabular flint, the Deepdale Flint, which rises westward from the foot of the cliff to form a clear marker in the lower part of the cliff face. Above are 2 deeply weathered notches formed by the Deepdale Lower and Upper Marls and the sequence can be traced upward to the Ravendale and Triple Tabular flint bands. These form equally clear markers at the head of the bay on the eastern cliff, the Ravendale Flint being up to 12 cm thick. The boundary of the *Terebratulina lata* and *Sternotaxis plana* zones lies at the base of the 50 cm thick unit of thinly bedded chalks directly below the Ravendale Flint, where the first *S. plana* occur. *Gibbithyris semiglobosa* is present just above the flint.

Once the Ravendale and Triple Tabular flints have been identified it is easy to retrace the succession downwards along the eastern cliff (Fig. 80) to the Melton Ross Marl, which forms a deep cleft in the cliff foot at low-water mark.

Locality 2. North Landing (Welton and Burnham Chalk formations).

From Thornwick Bay either walk along the cliff top path eastwards or drive back to the B1255 and turn left to North Landing (Fig. 75). The road terminates by a large pay car and coach park. From there walk down the cliff road adjacent to the café and pub. At the head of the bay on the left hand (western) side is a slight embayment and cave in the Chalk. A prominent ledge and crevice rising eastwards from beach level here marks the position of a useful marker bed, the Ulceby Marl. A second marker horizon can be picked up about halfway along the western side of the Landing, at a conspicuous marine arch. Here the

3 Excursions: Itinerary 12

Figure 80. Marker flint bands at the east side of Great Thornwick Bay. RF=Ravendale Flint, TT=Triple Tabular Flints.

Ravendale Flint is about 2 m above shore level and just above it are the Triple Tabular Flint bands, with the prominent Ludborough Flint about half way up the inside walls of the arch. At the north-western extremity of the Landing the remains of a ship's boilers lie in a deep cleft; a prominent rusty-brown flint band adjacent to them is the Ferruginous Flint. These three marker levels allow the whole succession (Fig. 76) to be followed, though care must be taken as much of the walk is over boulders and chalk ledges.

On the eastern side of the Landing the Beacon Hill Marl lies just above low water. A little higher in the sequence the thickest (20 cm) tabular flint, the Ludborough Flint, cuts across the mouth of Robin Lythes Hole, which leads into a magnificent cavern with another exit onto East Scar. The position of the Ulceby Marl is again marked by a ledge and crevice, which descends to the shore in the cliff immediately adjacent to the lifeboat slipway. At intervals of 2 m and 3.5 m above the marl two oyster beds are visible here, the higher representing the Ulceby Oyster Bed (Mortimore *et al.*, 2001).

Fossils are not common through most of the succession, but *Sternotaxis plana* occurs in the upper beds (from just below the Ravendale Flint). The lower of the two oyster beds is a bed of chalk about a 20 cm thick with very scattered, mainly fragmentary oysters (*Pycnodonte vesicularis*) and occasional brachiopods. The upper bed is similar but more shelly.

Both this and the following itinerary 13 occupy less than a full day but cannot be combined safely because by the time one is finished the tide will normally have risen too much for the other. Itinerary 15 can provide a short but interesting 'filler' to complete the day.

ITINERARY 13: FLAMBOROUGH HEAD

I.C. Starmer and P.F. Rawson

OS   1:25 000 Explorer 301 Scarborough, Bridlington & Flamborough Head
        1:50 000 Landranger 101 Scarborough
GS   1:50 000 Sheet 55/65 Flamborough and Bridlington

Locality 1. Selwicks Bay (structures in Chalk)

From Flamborough village follow the B1259 to Flamborough Head, where there is a large car park and café adjacent to the lighthouse (Fig. 75). Opposite the café there is a good view of Selwicks Bay. Here the chalk cliffs and foreshore are formed by the flint-bearing Burnham Chalk Formation, overlain by the flintless Flamborough Chalk Formation. The flinty chalk is referred to the lower *Hagenowia rostrata* Zone and the flintless chalk to the upper part of the same zone. Fossils are uncommon but occasional examples of *Gonioteuthis westfalica*, *Echinocorys* sp., *Porosphaera globularis*, small brachiopods and fragmented inoceramid bivalves are found. The inoceramids include *Cladoceramus undulatoplicatus* from the highest part of the Burnham Chalk, which is a basal Santonian marker in the German successions (Mortimore *et al.*, 2001, p. 426).

However, the main feature of geological interest is a zone of intense deformation running E-W through the bay, which represents part of the Howardian-Flamborough Fault Belt (Fig. 3). The zone shows a clear structural sequence of N-S compression (forming E-W folds and thrusts) followed by N-S tension (producing E-W striking extensional faults), both phases being related to the Cenozoic Alpine Orogeny. The structure has been described in detail by Starmer (1995a).

The bay is best investigated on a falling tide. From the cliff top, the view of the intertidal foreshore shows that it is a gentle, E-W syncline, with rocks in the south dipping gently north and those in the north dipping gently south, around a hinge which is usually visible at the edge of the beach (Fig. 81). Fractures represent later extensional faults which cut the synclinal structure, illustrating the sequence of N-S compression followed by N-S tension.

A path to the left of the lighthouse gate leads down the West Cliff to Selwicks Bay. On either side of the path, a variety of lime-loving flowers, including orchids, can be seen on the boulder clay slopes.

Viewing the cliffs from the beach, the sharpest features seen are the late extensional faults, striking E-W, but varying from steeply to gently dipping, with the upper sides (or 'hanging walls') moving downwards. The movements can be determined from slickenlines on the fault surfaces.

Starting in the south of West Cliff, the 5 m wide frontal fault zone (Fig. 82) is a series of sub-vertical E-W faults causing strong brecciation: this zone contrasts markedly with the immediately adjacent almost undisturbed Burnham Chalk on the south side. To the north of the frontal zone, subhorizontal to gently dipping extensional faults extend northwards to the new steps, where a steep extensional fault curves at beach level to dip more gently northwards beneath the cliff (i.e. it is a 'listric' fault, Fig. 83). On its northern side, in its 'hanging wall', the listric fault has cut earlier E-W folds and south-directed thrusts in a sheet which is limited at the top by a second, north-dipping listric fault. Northwards, the sheet of folds and thrusts disappears under the beach and the extensional faults become south-dipping along the rest of West Cliff: in other words, the north-dipping fan is re-

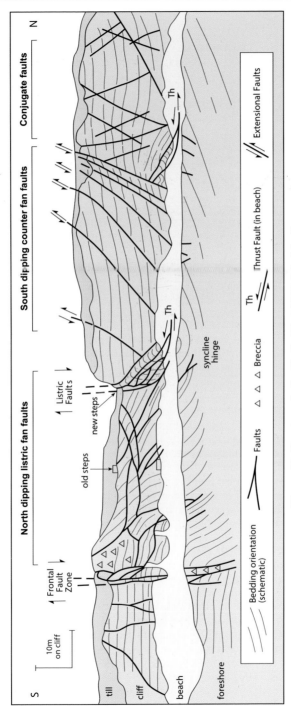

Figure 81. Structures in the Chalk, West Cliff, Selwicks Bay.

3 Excursions: Itinerary 13

Figure 82. West Cliff, Selwicks Bay, looking due west. Small bluff with vertical 'Frontal Fault Zone' on left (south) edge, sub-horizontal beds across the front, and steep northward-dipping beds and listric faults on the right (north) edge.

placed northwards by a south-dipping counter fan, in a similar way to structural developments in the North Sea.

About 35 to 40 m north of where the thrust and E-W folds disappear beneath the beach, they reappear again. This effect results from their gradual upthrow northwards on the 'footwalls' (undersides) of the later counter fan extensional faults, graphically illustrating the sequence of E-W folding and thrusting followed by extensional faulting. At the northern end of West Cliff, the features of the North Cliff start to dominate: these are NNE-SSW and NNW-SSE extensional faults, commonly forming crossing conjugate pairs.

The structure in West Cliff is not concordant with that in the foreshore beneath (Fig. 81), because of the thrust between them, beneath the beach. Considering the structures, the total downthrow across the whole Selwicks Bay zone is at least 18 m to the north.

At the north-east corner of the bay, a large re-entrant in North Cliff is called the Molk Hole and leads to several small caves and two spectacular arches. At the Molk Hole entrance, at the foot of the cliff, the highest flint band (the High Stacks Flint) of the Yorkshire Chalk is exposed, marking the top of the Burnham Formation. This location should be investigated only on a falling tide.

The south cliff of Selwicks Bay forms the northern side of Flamborough Head: all but the highest few metres consists of the Burnham Chalk Formation. This area lies south of the main zone of deformation and consequently faulting is less intense. High up in the cliff face at the furthest embayment, Common Hole, there is an overturned fold which may reflect tectonic movement but could be a much more recent feature: Lamplugh (1891b) thought it was formed by the drag effects of an ice sheet overriding the Chalk as it moved southward.

3 Excursions: Itinerary 13

Figure 83. West Cliff, Selwicks Bay. Near the new steps, at the left (south) end, a listric fault curves north round the base of the main cliff beneath a triangular zone containing older thrusts. Above and to the right (north), another north-dipping listric fault is succeeded by south-dipping counter fan faults. In the foreground, foreshore rocks dip around the syncline hinge at the left (south) edge of the photo.

Small embayments along the south cliff represent collapsed blow holes (note the concave chalk faces). The solitary sea stack is locally called 'Adam'; its former partner ('Eve') on the opposite side of the bay (on Kindle Scar) was illustrated by Lamplugh (1896, plate 31), but has since been eroded away.

The southern side of Flamborough Head is accessed by returning up the steps and walking around the lighthouse to the track leading to the Fog Siren.

Locality 2. High Stacks (viewpoint)

From the Fog Siren, follow the cliff top path south-eastwards to view High Stacks [TA 257 704], a chalk stack capped by glacial sediments (Fig. 84). The stack is separated from the main body of the chalk "by what may be the remnants of a palaeo-valley. Chalk clast rich gravel units at the base of the section may be either glacial outwash or possibly periglacial slope deposits. The lower half of the main exposure is composed of a grey col-

Figure 84. Quaternary deposits on Flamborough Formation chalks, High Stacks.

oured till, overlain by a significant thickness of gravels and sands, possibly with interbedded thin tills. The uppermost part of the section is composed of a relatively thin reddish brown (clearly weathered) till unit which Lamplugh (1891) regarded as the Skipsea Till of modern usage" (E.R. Connell, personal communication).

From High Stacks the chalk cliffs extend westward along the north side of Bridlington Bay to Sewerby Steps (a distance of about 6.5 km: Fig. 75), exposing a continuous 167 m succession of chalk representing all but the highest part of the Flamborough Chalk Formation. The sequence generally youngs westwards, although locally the gentle dips (0–15°) vary in direction, because of an underlying structure of gentle domes and basins on all scales, formed by the interaction of E-W and NNW-SSE folding (Starmer 1995a): however, in the western part of the section, the dips are dominantly south-westwards.

Unfortunately the former track leading down to the shore at High Stacks is no longer accessible and the next access point is about 3 km to the west, at South Landing (Itinerary 14). In between, about the lowest 26 m of the Flamborough Chalk Formation are exposed belonging to the South Landing Member. This belongs to the lower part of the upper *Hagenowia rostrata* Zone and consists of very hard, massive white chalk with a series of thin marls which increase in frequency up the succession to form a thinner-bedded sequence near South Landing.

ITINERARY 14: SOUTH LANDING TO SEWERBY

P.F. Rawson, E.R. Connell and I. Heppenstall

OS   1:25 000 Explorer 301 Scarborough, Bridlington & Flamborough Head
         1:50 000 Landranger 101 Scarborough
GS   1:50 000 Sheet 55/65 Flamborough and Bridlington

This itinerary focuses on both the Flamborough Chalk Formation (Fig. 85) and a series of Pleistocene sediments deposited in palaeo-valleys and banked against a buried interglacial cliff. The Chalk part of the itinerary follows an earlier version by the late Felix Whitham (in Rawson & Wright, 2000), revised, expanded and updated by P.F. Rawson. The important Quaternary sections are described by E.R. Connell and I. Heppenstall.

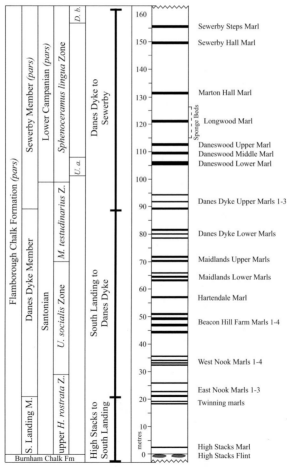

Figure 85. Lithic log of the Chalk from High Stacks to Sewerby Steps. Based on data in Whitham (1993), Mitchell (1994) and Mortimore *et al.* (2001).

From the B1259 at the south side of Flamborough village take a minor road signposted 'South Landing' at the crossroads [TA 228 702] (Fig. 75). From the Yorkshire Wildlife Trust Living Seas Centre at the car park a path leads down a ravine to the shore. On both sides of the Landing glacial deposits fill a palaeo-valley, some 250 m wide, cut into the chalk. This is one of a family of chalk dry valleys on the Yorkshire Wolds cut by periglacial fluvial systems during Pleistocene cold stages. Significantly, this valley form has preserved glacial sediments older than the Skipsea Till of the classic Holderness sequence allowing important new information on the chronology and extent of the last glaciation in the region to be revealed. This is in contrast to the Holderness coast to the south where the base of the glacial sequence is not seen and chalk bedrock is many metres below OD (Straw & Clayton, 1979 fig. 4.2). The modern stream at South Landing is excavating the glacial fill of this older valley.

Several years ago a storm cleared the beach in line with the present day ravine to reveal a chalk sequence distorted by folding or faulting, probably reflecting another deep-seated structure in the Howardian-Flamborough Fault Belt. This faulted ground may have formed a line of weakness exploited by periglacial and possibly glacifluvial erosion to form the palaeo-valley.

Locality 1. South Landing to Danes Dyke (Flamborough Chalk Formation and Pleistocene deposits)

To the east of the Landing the highest beds of the South Landing Member are exposed, where the tiny echinoid *Hagenowia anterior* is confined to about 4 m of accessible chalk (some 22 m above the base of the formation). It is possible to walk eastward from here working down the succession described by Whitham (1993).

1A. East Nook. The Pleistocene deposits above the chalk in the eastern cliffs (East Nook) at South Landing are poorly exposed due to slumping and the cliff is inaccessible without climbing aids. However, during the later years of the 19th century they were better exposed and were described by Lamplugh (1890). He recorded chalk bedrock declining in elevation to the north, defining the eastern flank of the palaeo-valley. Above chalk bedrock were both periglacially fractured chalk and a rubbly chalk gravel. This was overlain by what Lamplugh believed to be Basement Till containing sheared and deformed shelly clay, possibly equivalent to the 'Bridlington Crag' (rafts of shelly marine clay) previously known only from Bridlington north beach and Dimlington (Itinerary 17) in south Holderness. Above this unit Lamplugh recorded what is now termed Skipsea Till, of last glaciation date (see further details under 1B. West Nook).

The thick, coarse chalk gravels that Lamplugh recorded beneath the tills at East Nook/east cliff can be seen in the central area of South Landing where they reach over 10 m in thickness and chalk bedrock is not seen above modern beach level. These gravels are generally poorly bedded. The subangular-subrounded chalk clasts reach 0.30 m in size: 'exotic' erratics typically make up only about 5–6% of the 8–16 mm pebble fraction. In the main exposure bedding surfaces can be seen dipping at low angles to the west. The coarse texture of the deposit and poor rounding of clasts, in part reflecting debris flow deposition, indicate these beds were deposited in a high-energy proximal glacigenic fan environment with the ice sheet standing to the east of South Landing towards Flamborough Head. It is unclear if this unit was deposited during the Late Devensian glaciation or during an earlier event.

1B. West Nook. Lamplugh (1891) figured an important Pleistocene sequence close to West Nook which is often poorly exposed and has been overlooked since. Recent research has shown its importance to the understanding of the sequence and timing of the last glaciation in the area (Bateman *et al.*, 2015). The composite section is shown in Figure 86. All the deposits are thought to be younger than the thick chalk gravels seen in the central area of South Landing.

The deposits (Fig. 87) rest on a sub-horizontal chalk platform which is at a similar elevation (about 2.5 m O.D.) to the marine abrasion platform beneath the Ipswichian beach gravel at Sewerby (Loc. 3) and may be a remnant of the Ipswichian interglacial feature though here no marine sediments survive. The lowest unit consists of up to 4 m of well-imbricated chalk-rich gravels with interbedded, cross-bedded sands, recording flow from the north. 'Exotic' clasts (26–43%) comprise a range of igneous, metamorphic and sedimentary rocks of north British provenance. The upper 0.5 m of the gravel has been cryoturbated, indicating development of a periglacial landsurface after gravel deposition ceased. This erratic-rich gravel unit is interpreted as glaciofluvial outwash derived from an ice front to the north of the site. Presumably the ice sheet withdrew, deposition ceased and a periglacial landsurface subject to cryoturbation developed for an unknown period of time. Optically stimulated luminescence (OSL) dating of a sand bed within the gravels has given an age estimate of ~36.5 ka (Bateman *et al.*, 2015). Locally the upper part of this gravel unit has been calcreted with large blocks of this material present on the beach. It is unclear exactly when cementation occurred though it is likely to have been relatively recently.

Overlying the gravels are unfossiliferous laminated clays, silts and sands up to about 4.5 m thick, interpreted as a glacilacustrine deposit formed when drainage to the south in the palaeo-valley was blocked—presumably by the North Sea Lobe of the last ice sheet advancing some distance into what is now Bridlington Bay. The glacilacustrine facies are succeeded by a further 7 m of erratic-rich gravels and sands indicating renewed outwash deposition from the north, perhaps also signifying withdrawal of the icelobe in Bridlington Bay? OSL age estimates of 20.3 ka at the base of the glacilacustrine facies are matched by an age estimate of 20.0 ka near the top of the upper glacifluvial gravels and sands. These ages indicate rapid sedimentation rates for both of these units. Finally the ice sheet overtopped Flamborough Head and deposited a till which, based on stratigraphic position and matrix colour, correlates with the Skipsea Till at Sewerby (Loc. 3) and in the Holderness cliff sections (Itineraries 16 and 17).

The walk from South Landing westward to Sewerby Steps [TA 202 686] is approximately 3 km and exposes about 139 m of the Flamborough Formation sequence, from the higher part of the upper *Hagenowia rostrata* Zone (*c.* 6 m) through the whole of the *Uintacrinus socialis* (41 m) and *Marsupites testudinarius* (27.5 m) zones to the *Sphenoceramus lingua* Zone (just over 65 m exposed). Halfway along the section is the seaward end of Danes Dyke which is the only other exit from the beach before Sewerby. Care must be taken as high tide reaches the base of the cliffs in a number of places.

In contrast with the underlying formations the Flamborough Chalk is essentially flintless, and although there are many marl bands most are mere films. Thus there are far fewer obvious marker beds and although detailed sections have been measured (see logs in Whitham, 1993 and Mortimore *et al.*, 2001) and several of the thicker marls or marl and chalk complex bands named (Fig. 85) it is difficult to locate some of them during a field visit. The position of the more readily visible ones is indicated below.

Where the Flamborough Chalk reappears in the foot of the cliffs on the SW side of the Landing, a 2–3 cm thick marl, East Nook Marl 1 (East Nook Marl of Whitham,

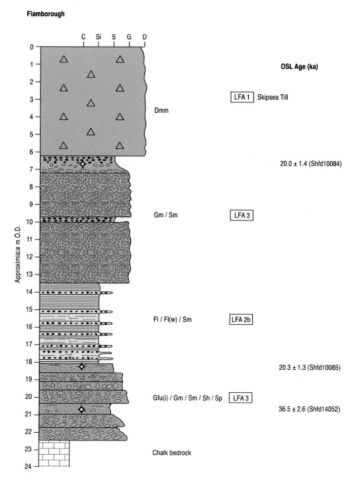

Figure 86. Composite section of the glacial deposits at West Nook, South Landing. Orange colours are gravels and sands, yellow is laminated silts, clays and sands, green is the overlying Skipsea Till. Optically stimulated luminescence (OSL) dates are shown to the right. From Bateman *et al.* (2015, fig. 6), reproduced by permission of the Geologists' Association.

1993), appears about 2.3 m above the cliff foot. This marks the base of the Danes Dyke Member, which is 66.7 m thick and extends from here to Danes Dyke. Compared with the South Landing Member, the lower to middle parts of the Danes Dyke Member are less hard, more thinly bedded and have more numerous, usually very thin, marl seams which are marked by narrow notches in the cliff face. Stylolitic layers (complex zigzag contacts formed by loading and solution processes) are common. In the middle part of the sequence there are numerous gently concave scours up to 2 m long (Mitchell, 1994). A number of minor faults dissect the succession.

The echinoid *Hagenowia anterior* is essentially confined to an 8 m sequence spanning the boundary of the Danes Dyke and South Landing members and occurring either side of South Landing. To the west of the ravine the horizon can be traced for about 100 m

3 Excursions: Itinerary 14

Figure 87. Lowest part of the glacial succession at West Nook recorded in Figure 86. Pebble-cobble chalk gravels rest on a sub-horizontal chalk platform. Note the locally deformed upper part of the gravel with vertical clast orientation—a subaerial periglacial land surface. Junction with overlying deposits is obscured here by slumped deposits and turf. Walking stick (1 m) for scale.

before dipping below beach level. At other levels this species is fairly rare: a few specimens are recorded elsewhere in the *rostrata* Zone and there are isolated examples in the lower *Uintacrinus socialis* Zone and the *Sphenoceramus lingua* Zone. The remainder of the fauna in this part of the *rostrata* chalk is restricted to rare brachiopods, fragmented inoceramid shells, thick shelled *Echinocorys* sp., the belemnites *Gonioteuthis westfalica granulata* and *Actinocamax verus*, and sponges (including *Amphithelion*, small varieties of *Laosciadia plana, Siphonia koenigi, Stichophyma tumidum* and abundant *Porosphaera globularis*).

1C. Beacon Hill. Plates of the zonal index *Uintacrinus socialis* first appear about 4 m below the lowest of the four West Nook Marls, some 360 m west of South Landing. The overlying Beacon Hill Farm Marls 1–4 [TA 226 692] lie below Beacon Hill, about 600 m from the Landing: they occur through a 5.8 m sequence and are the thickest marls in the South Landing Member, the lowest being about 8 cm thick, the others 3-5 cm thick. They are visible in the lowest part of the cliff in the area where the cliff face marked by three broad bands of brown downwash and staining. About 300 m further along, and hidden from sight until one is almost upon it, is Hartendale Gutter—the floor of the gutter being about 7 m above shore level. On the east side of the gutter the Hartendale Marl is visible 2–2.5 m above shore level.

This part of the sequence was one of the very few sections in England to yield complete cups of *Uintacrinus socialis*. It is now almost impossible to find complete specimens in the lower, accessible parts of the cliffs, but isolated plates are common and at some horizons small groups of 10 or more plates occur, possibly all forming part of the same individual. Other fossils in the *socialis* Zone include fragmented inoceramids, *Echinocorys* spp., *Orbirhynchia pisiformis*, *Pseudoperna boucheroni*, *Parasmilia* sp. and sponges belonging to the same group as listed for the previous zone. Belemnites of the *Gonioteuthis granulata* lineage appear to be less common and the occurrence of *Uintacrinus socialis* plates diminishes in the higher beds, being very rare in the highest 3 m of the zone.

The 3 Maidlands Lower Marls appear about 200 m west of Hartendale Gutter and just before a small headland, at TA 219 692. Approximately 100 m further along and about 220 m before Danes Dyke is reached, rare isolated plates of *Marsupites testudinarius* appear, the first 0.7 m below Maidlands Upper Marl 1 (Mitchell, 1994) [TA 218 692]. This horizon marks the base of the *Marsupites testudinarius* Zone, which reaches its maximum thickness in this area. Massive bedding continues upward from the higher part of the preceding zone but the chalk becomes softer. The lowest 15 m of the zone occur in the cliff up to Danes Dyke, but *Marsupites* plates are generally very rare here, though Mitchell (1994) recorded two flood occurrences in this part of the sequence. In the cliff immediately east of Danes Dyke the three 2–3 cm thick, closely spaced Danes Dyke Lower Marls form a clear marker (Fig. 88), rising from the shore into the lower cliff.

It is possible to return up the cliff at Danes Dyke, following a footpath inland to Flamborough village. There is also a car park here.

Figure 88. The three Danes Dyke Lower Marls, 50 m east of Danes Dyke. These form a clear marker in the upper part of the Danes Dyke Member.

Locality 2. Danes Dyke to Sewerby Steps (Flamborough Chalk Formation and Pleistocene deposits)

2A. Danes Dyke. The seaward end of the ravine at Danes Dyke/Dykes End again exposes a palaeo-valley, here cut into *testudinarius* Zone chalks. It is smaller than that at South Landing, only some 150 m in width. This valley too is filled with a complex succession of periglacial and glacial sediments, figured by Dakyns (1880) and Lamplugh (1891). These deposits have been re-excavated both by the modern stream flowing in the ravine and by Bronze Age people constructing an entrenchment and bank across Flamborough Head. The Chalk is folded and faulted here, reflecting reactivation of the E-W Langtoft Fault (Starmer, 2013): the structures are particularly visible when the intertidal zone is exposed. The occurrence of faults and folds in the chalk probably facilitated periglacial erosion of the palaeo-valley during Pleistocene cold stages.

The succession is well exposed in the west cliff (Fig. 89), where over 30 m of sediments lie on tectonically and periglacially brecciated chalk bedrock. Above this is a wedge of westerly thickening gravels, up to 3 m thick. The clasts are exclusively of chalk derived by frost-shattering of the valley side before being soliflucted/geliflucted down the slopes. However, a cluster of ?Jurassic sandstone boulders at the base of this deposit suggests they are the remnant of older glacial sediments within the palaeo-valley. The thin silt laminae in the unit and matrix silt of the main deposit are believed to be aeolian loess derived by deflation of outwash deposits in front of the southward advancing North Sea Lobe of the last ice sheet. Very similar periglacial facies are present nearby at Sewerby buried cliff (Loc. 3) and at other inland sites in East Yorkshire (Catt *et al.*, 1974). Interestingly periglacial sediments have not been preserved in the South Landing palaeo-valley.

Figure 89. Quaternary deposits in west cliff, Danes Dyke as exposed in 2009. The visible section is some 18 m high; the cliffs adjacent to the exposure rise to approximately 35 m OD. The spade to the left of the rucksack is 1.08 m long.

Above the periglacial slope deposit is a unit of laminated silt, clay and fine-medium sand, 1 m thick and thinning to the west. This was deposited in a glacial lake. As at South Landing, the presence of a glacilacustrine facies in a south-oriented valley suggests ponding by the nearby ice sheet having advanced into the northern part of Bridlington Bay. These sediments are erosively overlain by the first of at least three till units in the exposure. Between each till are thick units of erratic-rich gravels and sands deposited either as proglacial outwash or subglacially. The first and third tills have a matrix colour resembling the Skipsea Till of Holderness, while the second till is greyer, more like the Basement Till—which Lamplugh (1891) recorded from this site. However, ongoing research indicates that the entire complex dates to the Late Devensian glaciation, being deposited by the advance of the North Sea Lobe (Bateman *et al.*, 2015).

The 1.7 km length of chalk cliffs and adjacent scars from the western side of Danes Dyke to Sewerby buried cliff are formed by the 71.5 m thick Sewerby Member. Its appearance represents a lithological change from the more massive chalks of the upper part of the Danes Dyke Member to thinner-bedded sequences with many thin marl partings and eight thicker, named seams. Scours occur in the lower part and at various levels higher in the sequence. The base of the member is formed by a 3–4 cm marl, the Danes Dyke Upper Marl 1, which appears at the foot of the chalk cliff immediately west of Danes Dyke [TA 215 691]. The member embraces the remaining 11 m of *Marsupites testudinarius* Zone chalk and just over 60 m of *Sphenoceramus lingua* Zone chalk.

About 100 m along the section *Marsupites* plates become extremely common and complete calyces (cups) are found. Floods of the zonal species are spread over about 5 m of chalk in the middle part of the zone (Whitham, 1993; Mitchell, 1994) and it then dies out close to the upper boundary about 200 m SW of the dyke. The higher part of the zone is characterised by shell-detrital chalks composed of both fragmentary and complete oysters and inoceramids—a local representative of the widespread north European Grobkreide (coarse chalk) facies (Mitchell, 1994; Mortimore *et al.*, 2001). Other species occurring in the upper part of the zone, west of the dyke, include *Pseudoperna boucheroni* (in bands), *Orbirhynchia pisiformis*, occasional *Echinocorys* sp., and *Ventriculites*, *Porosphaera globularis* and other sponges often preserved as oxide films. Large examples of *Gonioteuthis granulata* occur.

Near to four seaweed-covered calcreted gravel blocks (first noted by Rowe, 1904), the base of the *Sphenoceramus lingua* Zone is marked by a profusion of fragmented shells of the zonal species and it is at this level that the last *Marsupites* occurs (Whitham, 1993). This horizon also marks the boundary of the Santonian and Campanian stages. About 6 m higher in the section is the 9 cm Daneswood Lower Marl, which comes down to beach level about 25 m west of the calcrete blocks, at TA 213 690. Some 25 m further west the 4–5 cm Daneswood Middle Marl reaches the shore.

Chalks of the *lingua* Zone are mainly fairly hard with some massive bedding interspersed with a series of thinly bedded horizons. Stylolitic surfaces are less frequent than in the previous zone. Fossils are common at some horizons with *Sphenoceramus lingua* the dominant bivalve while rare examples of *Sphenoceramus pinniformis* occur. Mitchell (1994) recorded *Uintacrinus anglicus* over a 9 m interval just above the base of the *lingua* Zone and extended Bailey *et al.*'s (1983) *Uintacrinus anglicus* Zone to this interval: it is regarded here as a subzone.

Sponges are far more common in the *lingua* Zone than elsewhere, with the best developed concentration of hexactinellid and lithistid sponges occurring in the famous Flamborough Sponge Beds, which consist of just over 10 m of chalk, the basal beds lying some 15.5 m above the base of the zone and about 1.5 m above the 5 cm Daneswood

Upper Marl—which reaches the shore close to two calcrete blocks and some 35 m west of the Middle Marl. The more shallow dip of the strata where the Sponge Beds reach the shore provides a continuous exposure on the beach scars for a considerable distance, commencing about 350 m from Danes Dyke; the main exposure lies nearer to this ravine than to Sewerby. In the middle of the Sponge Beds is a 6 cm grey marl, the Longwood Marl. Many fine sponges occur in both cliff and scars, including *Pachinion scriptum*, *Stichophyma tumidum*, varieties of *Laosciadia plana*, *Siphonia koenegi*, *Rhizopoterion cribosum*, *Amphithelion* (*Verruculina*), *Wollemania laevis*, *Sporadoscinia* strips, *Leiostracosia punctata* and *Porosphaera globularis*. Also occurring in the Sponge Beds are very large *Echinocorys* (up to 80 mm long), *Sphenoceramus lingua* and *Sphenoceramus pinniformis*, with rare *Gonioteuthis granulata*. The top of the Sponge Beds is marked by three thinly bedded 20 cm chalk horizons spread over 1.5 m, with the intervening beds containing abundant *Pseudoperna boucheroni*.

Other species recorded in the *lingua* Zone include a band of *Offaster pilula*, *Hagenowia* sp., large and rounded forms of *Cardiotaxis*, *Hypoxytoma tenuicostata*, *Orbirhynchia* sp. and the rare ammonites *Hauericeras pseudogardeni* and *Scaphites* sp. Echinoid spines and asteroid plates are common in the lower half of the zone.

2B. Sewerby Steps. Above the Sponge Beds, towards Sewerby Steps, the more massive bedded chalk becomes less fossiliferous, with sporadic occurrences of *Echinocorys*, sponges, fragmented inoceramids and shell debris. A single specimen of the ammonite *Discoscaphites binodosus*, index species of the *binodosus* Subzone, indicates that the top 4 m of chalk at Sewerby Steps belongs to the lowest part of this subzone: it is much better developed inland (Whitham, 1993).

Locality 3. Sewerby Buried Cliff (Pleistocene features)

From Sewerby Steps the highest beds of the Sewerby Member continue for about 300 m SSW before the modern cliff face shows the chalk terminating abruptly against interglacial beach shingle, blown sand, periglacial 'head' and glacial tills (Fig. 90). This is the famous Sewerby Buried Cliff section, the interface marking a Pleistocene cliff face carved in the Chalk, which runs slightly obliquely to the modern Chalk cliff and can be traced for at least 50 metres before disappearing completely behind the glacial deposits. The buried cliff then strikes inland to run along the dip slope of the Chalk Wolds to the Humber (where it was visible at Hessle) and on into Lincolnshire. It is an interesting and important feature in the glacial chronology of the area (Catt & Penny, 1966). At the foot of the buried cliff is an interglacial shingle beach ('Sewerby Raised Beach'), about 1 metre above the modern beach level and resting on a planed surface of chalk and the Basement Till, though the contact with the till is only seen on the rare occasions when storms have stripped off the modern beach. The raised beach contains vertebrate remains indicative of the Last (Ipswichian) Interglacial, including the straight-tusked elephant (*Palaeoloxodon antiquus)*, the narrow-nosed rhinoceras (*Dicerorhinus hemitoethus*) and *Hippopotamus*. It dates to between 116,000 and 128,000 years ago.

The periglacial deposits (blown sand and periglacial 'head' slope deposits derived from chalk bedrock) above the raised beach mark the early phase of the encroaching cold period of the Devensian (last) glaciation. As the climate deteriorated further ice spread over the area some 20,000 years ago to deposit the Skipsea Till, which blanketed the whole sequence. Over most of Holderness the Skipsea Till is believed to rest directly on Basement Till. At Dimlington, south Holderness (Itinerary 17) the two tills are separated by

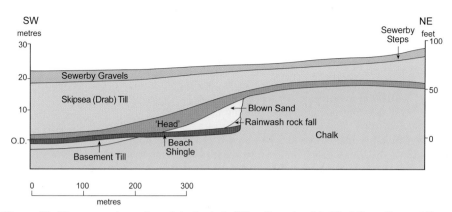

Figure 90. Diagrammatic section of the buried cliff at Sewerby. Modified from Catt and Penny (1966, pl. 24).

the Dimlington Silts which have been both radiocarbon and OSL dated to the Upper Devensian (about 20.9 ka: Table 6). The Sewerby succession is thus crucial in demonstrating that the tills represent deposition during two different glacial periods separated by a warm interglacial. The Basement Till probably represents the 'Wolstonian' (penultimate) glaciation, about 140,000 years ago (Table 6 and Catt, 2007).

note: If the whole of the South Landing to Sewerby section is followed on a falling tide there will still not be time to return along the shore. Instead either go into Sewerby village and catch a bus back to Flamborough, or return along a footpath which follows the cliff top to Danes Dyke and then strikes inland to Flamborough.

ITINERARY 15: LANGTOFT, FOXHOLES AND STAXTON HILL

P.F. Rawson

OS 1:25 000 Explorer 301 Scarborough, Bridlington & Flamborough Head.
    1:50 000 Landranger 101 Scarborough
GS 1:50 000 Sheets 54 Scarborough, 64 Great Driffield

This brief itinerary links three localities (Locs 2–4) along the B1249, which runs northwards across the Wolds from Driffield to Staxton. Two show inland exposures of chalk shatter zones in disused quarries immediately by the roadside, next to which it is possible to park. The last locality, Staxton Hill, is an excellent viewpoint from which the glacial and post-glacial history of the Vale of Pickering area can be demonstrated. Combined with the Flamborough area itineraries 13 or 14 this can make a full field day, in which case drive from Flamborough to the west side of Bridlington and turn off the A165 onto the B1253 through Rudston.

Locality 1. Rudston churchyard

The churchyard at Rudston village [TA 098 678] contains the Rudston monolith, reputed to be the tallest standing stone in England. Dating to the Bronze Age, it is made of Jurassic sandstone and must have been transported for at least 16 km, possibly from the Pickering area (Allison, 1976). The grave of the East Riding's best-known novelist, Winifred Holtby, lies near the SW corner of the churchyard.

Locality 2. Langtoft (Chalk structures)

Continue westwards from Rudston to a traffic island at Octon Cross Roads where the B1253 crosses the B1249. Turn left (southwards) along the latter and drive through Langtoft. Just south of the village on the east side of the road [TA 012 659] is a disused chalk quarry with a small tarmacked parking area outside the gate. Here, part of the Flamborough Chalk Formation (probably the *rostrata* Zone) is exposed. The chalk at the southern (right-hand) end of the section is almost horizontal, but northwards it is dragged up to dip of about 50°, before passing into a zone of brecciated, calcite-veined and slickensided chalk beneath the grassy slopes at the northern end of the quarry. Only one small patch of this brecciated chalk is now visible. Note that in a quarry across the dale from here the chalk is almost horizontal, but listric shears occur, dipping 45–70° WSW (Starmer, 1995a). Starmer has suggested that this 'shatter zone' probably reflects post-alpine extensional reactivation of the E-W trending Langtoft Fault, which can traced beneath the Chalk eastwards along the northern side of Bridlington and into the offshore area (Kirby & Swallow, 1987).

Locality 3. Foxholes (Chalk structures)

From Langtoft head northwards to Foxholes. About half a kilometre north of the village on the east side of the road [TA 012 735] it is possible to park on a small patch of rough ground in front of another old quarry. This exposes flinty chalk of the Burnham Chalk Formation. The main face shows a mass of chalk folded to dip more or less uniformly north at about 70°, whereas in the top right-hand corner of the quarry and along the right-hand side, the chalk is almost horizontal. Unfortunately this area and the faulted contact between dipping and horizontal chalks are now almost completely covered in vegetation. Starmer (1995a) has shown that the folding represents compression of probable Alpine age, while the fault marks a later tensional phase. This 'shatter zone' again appears to lie over a pre-Chalk fault and extends eastwards to merge with the Bempton zone on the coast at Staple Nook (illustrated in Starmer, 1995b).

Locality 4. Staxton Hill (viewpoint; glacial and post-glacial features)

Continue northward to the top of Staxton Hill [TA 009 778] and park at the picnic area and viewpoint (signposted). There are public toilets here and tables in the picnic area. On a clear day there is a spectacular view of virtually the whole Vale of Pickering. In the immediate foreground the scarp edge is formed of chalk underlain by the Hunstanton and Speeton Clay formations, while much of the Vale is floored by the Kimmeridge Clay Formation. The Corallian dip slope of the Tabular Hills rises away from the observer in the distance. The main reason for stopping here is to consider the Late-Glacial and immediately post-glacial history of the area. The following summary is based mainly on the late Dr L.F. Penny's account (in previous editions of this guide) and a recent review by Evans *et al*. (2017).

Directly opposite the viewpoint, on the far side of the vale, the Wykeham lakes are clearly visible: these fill former quarries adjacent to the Wykeham Moraine, which originally curved halfway across the Vale of Pickering (Fig. 91). The moraine is generally believed to mark the westerly limit of penetration of the North Sea ice lobe (Penny & Rawson, 1969), which blocked the eastern end of the Vale of Pickering, while the other end of the vale was simultaneously blocked by Vale of York ice at the Coxwold-Gilling and Kirkham gaps. Glacial Lake Pickering extended between the two ice lobes, depositing lacustrine clays which have been proved in numerous boreholes. The lake reached a maximum elevation of about 70 m above sea level at or before about 21 ka. It was fed principally by the waters of the Newtondale spillway which deposited the fan on which Pickering stands, and also by those of the Forge Valley. The latter, hemmed in between the North Sea ice lobe at Wykeham and the Corallian dip slope, were forced westward, depositing the Hutton Buscel kame terrace and the delta fan which spread south-westward from the point where it entered the lake.

As the North Sea ice retreated from the Wykeham moraine, the Seamer-Scarborough Valley was uncovered, which is the prominent steep-sided valley on the skyline some distance to the right (east) of the Wykeham lakes. Water from this valley then similarly flowed into the lake and deposited the Seamer delta. By around 17.6 ka the lake level had fallen to about 45 m, possibly as water could escape via the Kirkham Gap. As North Sea ice retreated further east, Lake Pickering fell further to 30 m, being impounded not by ice but against the Flamborough moraine. Eventually the lake breached this at a spillway found at Filey.

Lake Pickering probably remained in a diminished form and at a lower level until the Kirkham spillway had been finally cut down to its present level, though a relic remained in post-Glacial times in the form of Palaeolake Flixton. From Staxton viewpoint the site of this lake lies to the extreme right on the far side of where the A46 can be seen crossing the vale. The lake lay between the Wykeham and Flamborough moraines and included small peninsulas and islands; its sedimentary history and palaeogeography varied through time (Palmer *et al*., 2015). The shores of this lake provided favourable settlement sites for

3 Excursions: Itinerary 15

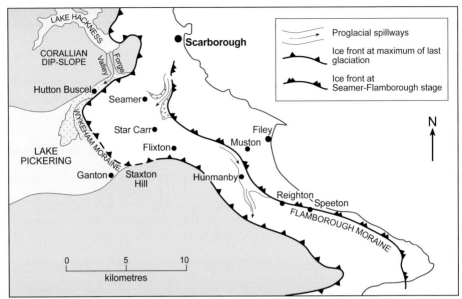

Figure 91. Glaciology of the eastern end of the Vale of Pickering. Redrawn from Penny and Rawson (1969, fig. 3), by permission of the Geologists' Association.

Mesolithic man around Flixton and Starr Carr, which form the most important Mesolithic site in Britain (e.g. Clark, 1954; Milner *et al.*, 2012).

With the demise of Lake Pickering the waters which had been diverted through the Forge Valley were able to flow straight into the Vale of Pickering initiating the present course of the Derwent and destroying the southern half of the Wykeham moraine (which probably abutted on the Chalk scarp around Ganton). The valley was still fen in historical times, of which the many 'carrs' (fens) and 'ings' (water meadows) bear witness throughout the area, and is still liable to severe flooding despite the cutting of the Hertford River and the canalisation of the Derwent in the eighteenth and nineteenth centuries.

3 Excursions: Itinerary 16

ITINERARY 16: NORTH HOLDERNESS: BARMSTON SOUTH TO MAPPLETON

E.R. Connell

OS   1:25 000 Explorer 295 Bridlington, Driffield & Hornsea,
           Explorer 292 Withernsea and Spurn Head
       1:50 000 Landranger 107 Kingston upon Hull, Beverley & Driffield
GS   1:50 000 Sheet 55/65 Flamborough and Bridlington

At the Danes' Dyke and South Landing sites (Itinerary 14) on Flamborough Head it is possible to see glacial and periglacial sediments deposited before the advance of the last ice sheet (during the Late Devensian Dimlington Stadial approximately 26–15,000 radiocarbon years ago) overtopped Flamborough Head and advanced into what is now Holderness. Both in north and south Holderness (Itinerary 17) the coastal cliff sections display sediments deposited as the ice sheet withdrew to the east across Holderness creating the low but complex and irregular morainic topography.

Locality 1. Barmston cliffs (Skipsea Till and proglacial lake sediments)

Barmston village lies immediately east of the A165, 7 km south of Bridlington. Drive east through the village to the car park to the right of the road at the cliff top [TA 170 593]. A parking charge applies here and tickets can be obtained from the caravan park reception across the road. Having parked and paid walk back into the caravan park to reach the cliff-top path and walk north for about 700 m until the path descends towards the beach at approximately TA 170 600.

Once on the beach walk south until the cliff section seen in Figure 92 is visible. The range of sediments deposited as the ice sheet withdrew can be seen by walking south from this exposure to a position below the road end car park. There are steps cut into the till cliff just south of this point to reach the cliff top and the car park. If the steps have been eroded walk back to the north along the beach and retrace your steps up to the car park.

Dark greyish brown Skipsea Till forms the base of the exposures (Fig. 92). Whilst generally massive looking, matrix-rich and with few large clasts, careful observation reveals a range of bedding/lamination together with small-scale fold and shear structures which together with striated clasts indicate deposition beneath the overriding ice sheet. Walking south the upper surface of the till is irregular—sometimes at beach level, other times with up to 3–4 m exposed in the adjacent cliff. It is unclear if this geometry is due to slumping and settlement, or perhaps subtle proglacial deformation by a re-advancing ice margin. At some points along the section 'pendent structures' can be seen where gravels have loaded and sunk into the water-saturated upper portion of the till soon after deposition. Erratic clasts in the till come from bedrock sources to the north and include Cretaceous, Jurassic, Permo-Triassic and Carboniferous sediments together with a range of igneous and metamorphic rocks from northern England and Scotland and, much more rarely, Norway.

The irregular upper surface of the till is draped by up to 2 m of rhythmically laminated silts and clays formed as bottom-set sediments in a proglacial lake as the ice sheet withdrew. They are believed to be varves, annually deposited sediments, suggesting the lake may have been in existence for at least 350 years. They grade up into ripple-laminated, and locally load-deformed sands and finally high-angle foreset beds (Fig. 92) recording the advance of a delta front, fed by meltwaters from the ice sheet into the proglacial lake.

3 Excursions: Itinerary 16

Figure 92. The northern end of the Barmston section. Skipsea Till (bottom left) is draped by laminated silts and clays (to the right) of a proglacial lake and the inclined sand foresets of a prograding delta. The section is approximately 6 m high.

At a number of points along the section wedge-shaped structures can be seen penetrating downwards from close to the ground surface. These are ice-wedge casts recording the establishment of permafrost conditions after the delta was abandoned. Recent research at Barmston (Bateman *et al.*, 2015) has provided a range of optically stimulated luminesence (OSL) dates indicating that the Skipsea Till was deposited after about 21.5 ka, with the proglacial lake sediments being deposited over the period from 15 ka to 11.26 ka as the ice sheet withdrew to the east and north.

Locality 2. Withow Gap, Skipsea (Skipsea Till and lake margin sediments)

Drive back west from the Barmston car park and rejoin the A165, continue for about 800 m south-west and turn left at Lisset on to the B1242. Continue through Ulrome and Skipsea villages (some 4 km) until reaching Southfield House on the road south from Skipsea village [TA 175 546]. Turn left into the private customer car park of Mr Moo's Ice Cream Parlour and Coffee Shop, where refreshments can be obtained. Park here and follow the marked path for about 800 m down to the low cliffs at Withow Gap [TA 184 546]. Take care in accessing the beach as coastal erosion can expose slippery muds.

The path down to Withow Gap passes between areas of relatively high morainic topography until a level area is reached some 300 m from the cliff section. This is an extension of the lacustrine sediments exposed in the cliff. At Withow Gap itself sediments are exposed recording deposition at the margins of a once more-extensive freshwater mere (Fig. 93). When John Phillips recorded the site in 1829 the exposed section was "about a quarter of a mile" (400 m); it is now reduced to about 150 m along this rapidly eroding stretch of coast. Meres, once widespread in Holderness, are areas of open water ponded on the irregular morainic topography. Many have been drained artificially, others breached and drained by coastal erosion. Organic deposits of former meres can be found along the Holderness coast, both beneath modern beach sediments (recorded on OS maps as Submarine Forest) or preserved along cliff tops. Hornsea Mere (some 8 km

3 Excursions: Itinerary 16

Figure 93. The northern exposure of the mere deposits at Skipsea Withow Gap. Skipsea Till overlain by grey muds, peat containing tree trunks and finally brown slope wash and soil. Spade is 1.08 m.

south of Skipsea) is the only one now remaining. A significant amount of research on palaeo-environments during the Late Devensian Late Glacial into the Holocene has been undertaken over many years at the important Withow Gap site, and other former meres in Holderness (Gilbertson, 1984).

The surviving lake-margin sediments at Withow Gap rest in a depression on the Skipsea Till (Fig. 93). Overlying the till are over 1 m of blue-grey massive to laminated muds with local beds of silty detritus peat. Plant debris and shells can be found (including the gastropod *Valvata piscinalis*). Augering has proved a sequence in places over 3 m thick, and the beds may occasionally be seen lower down the beach when storms have cleared the sand. At both the south and north ends of the site gravels and sands can be seen representing lake margin delta-like bodies. It is likely that the skull and antlers of the giant deer (*Megalocerus giganteus*) recorded by Phillips (1829) came from one of these early deposits, together with the remains of reindeer, red deer and aurochs. Radiocarbon dating of a birch log indicate ages older than 10,440 ±50 C14 years for these early deposits formed in the variably cold climate of the Devensian Late Glacial.

Unconformably overlying these sediments is a series of silty and variably humified peats up to 2 m thick in present exposures. This unconformity may relate to lake-level fluctuations driven by changing climate. The peats become less silty and richer in wood remains (mainly oak (*Quercus*), hazel (*Corylus*) and alder (*Alnus*)) upwards. Mesolithic flint tools have been recovered from these sediments and some wood fragments may record evidence of early woodworking. Pollen analysis of the peats, and earlier sediments, have revealed the changing woodland structure and climate through time. The combined evidence points to these peats recording an oak/alder carr environment on the margins of the mere with dates of 9880±80 and 4500±50 C14 years bracketing its development and spanning early to mid Holocene time. The uppermost 0.5–1.0 m of sediment in the section are grey-brown sandy silts in which is developed the modern soil profile together with some made ground.

Locality 3. Mappleton (Skipsea Till)

From Withow Gap drive south along the B1246 through Hornsea (about 8 km) to Mappleton village, 3 km further south. In the village turn left just past the garage and with the church on the right. Drive a few hundred metres to the cliff-top (free) car park [TA 228 438].

From south of Skipsea the coastal cliffs get higher and are more prone to collapse and slumping so care is needed if approaching the cliffs closely. Walk down the ramp on to the beach: sections are similar to both the north and south. These cliffs record thicker sections (up to 10–15 m) in the Skipsea Till, though the base is not seen. Unlike in the Flamborough, Barmston and Withow Gap sections, here the Skipsea Till shows significant variation in matrix colour (Fig. 94). The lowest 2.5 m visible are dark brown in colour (similar to the younger Withernsea Till: see Itinerary 17 South Holderness) overlain by 0.6 m of reddish brown till. This latter unit seems to have been recognised first by Reid (1885), who named it the 'red band'. North (right) dipping shear planes can be seen within these lower units indicating ice flow from that direction. Above the 'red band' the dark greyish-brown matrix colour is more typical of the Skipsea Till. There may be a thin representative of the Withernsea Till at the top of the section though this could be weathered Skipsea

Figure 94. Matrix colour variation within the Skipsea Till. The section is north of Mappleton [at TA 220 453] and was recorded in October 2016. It is approximately 10 m high. The spade is 1.08m.

Till. W.S. Bisat, in his research on the glacial sediments of the Holderness coast cliffs during the 1930s and 40s, also recorded this type of succession. He named the till beneath the 'red band' the Middle Drab, that above the Upper Drab (Bisat, 1962), all now referred to as Skipsea Till. This vertical matrix colour variation has recently been recorded from north of Hornsea to the Aldbrough area, a distance of about 10 km. It appears to reflect a vertical provenance change within the Skipsea Till reflecting incorporation of more Permo-Triassic sediment in the lower facies than in the overlying dark greyish-brown facies. In turn this is likely to reflect changing flow paths through time of component parts of the North Sea Lobe during the Late Devensian glaciation.

ITINERARY 17: SOUTH HOLDERNESS: WITHERNSEA AND DIMLINGTON

E.R. Connell

OS   1:25 000 Explorer 292 Withernsea & Spurn Head
          1:50 000 Landranger 107 Kingston upon Hull, Beverley & Driffield
GS   1:50 000 Sheet 81 Patrington

In this itinerary all three of the Holderness tills can be seen, together with the Dimlington Silts which rest on the Basement Till. At depth, the gently east-dipping Chalk platform beneath Holderness is cut by deep palaeo-valleys focused on Withernsea (Straw & Clayton, 1979 fig. 4.2; Berridge & Pattison, 1994 fig. 21). These may be old courses of the River Humber and a river flowing north-east from the Kirmington area in north Lincolnshire, probably cut during cold stages of the earlier Pleistocene when contemporary sea levels were low. Chalk is encountered in boreholes beneath Withersea town at between -34 and -38 m OD. These boreholes also confirm the presence of the Skipsea and Basement Tills (Loc. 2) beneath the Withernsea Till in the area (Berridge & Pattison, 1994 fig. 23).

Locality 1. Withernsea north cliff (Withernea Till)

In the northern outskirts of Withersea park in the car park at TA 336 287, or on the adjacent side road. Walk the short distance to the northern sea defences and descend the concrete steps to the beach, then walk north along the beach for about 400 m. In contrast to the sites in northern Holderness the cliffs here are composed entirely of the younger Withernsea Till (Fig. 95). Whilst present at the cliff top in the Mappleton area this till unit thickens southward. The older Skipsea Till passes below beach level at Tunstall (north of Withernsea) and does not rise into the cliff base again until near Dimlington (Loc. 2).

Up to about 10 m of Withernsea Till can be seen in the north cliffs dependent on their state. Locally, slumping obscures detail. The Withernsea Till is again matrix rich with a typically dark brown colour, though subtle colour variations are seen and attenuated lens of soft bedrock lithologies are often present (e.g. the deformed grey band in Fig. 95). Whilst the bedding appears massive, locally more complex finer bedding can be seen. For example, in the exposure shown in Fig. 94 the basal part of a till bed appears bedded but passes up into more massive till though with a grey folded and thrust lens. Deformation to the left of the large clast in the figure demonstrates ice flow from the right (north). There are few chalk clasts within the Withernsea Till: the majority of erratics are Jurassic and Carboniferous sedimentary rocks. Intra-beds of sands and gravels within the till may be exposed and large, striated, boulders of Carboniferous limestone can be conspicuous in the sections.

Locality 2. Dimlington (Basement Till, Dimlington Silts, Skipsea and Withernsea tills)

Leave Withernsea travelling south on the A1033 but turn left onto the un-numbered Holmpton Road in the southern outskirts of the town. Follow this road through Holmpton and Out Newton villages, a distance of approximately 9 km. On reaching the Easington Gas Terminal as the road swings to the south look for a turn into the left at TA 399 203. Park here and walk down the ramp to the beach. Care is needed as erosion can often steepen the lower part of the ramp. Walk north along the beach observing the sections as far as Dimlington High Land [TA 391 214], a distance of about 1.5 km. The cliffs here

3 Excursions: Itinerary 17

Figure 95. Withernsea Till exposed in Withernsea north cliff. The exposure [at TA 333 293] is 2 m high.

reach a height of 30 m OD, the highest in Holderness, and are prone to slumping. If the cliffs are in a poor state when visited there are excellent photographs of sections in the BGS memoir (Berridge & Pattison, 1994; these photographs are also available to view on the BGS website: http://geoscenic.bgs.ac.uk search 'Dimlington').

The complex Pleistocene stratigraphy in the Dimlington cliffs has been studied since the 1820s, and W.S. Bisat's work in the 1930s and 1940s clearly demonstrated the importance of the site for the chronology of glaciation in eastern England. It is one of only a small number of sites where the Basement Till can be seen in section. In 1969 radiocarbon dates were published from the Dimlington Silts (see below) of approximately 18,240–18,500 C14 years indicating that the overlying Skipsea and Withernsea Tills were deposited by the ice sheet close to the end of the last cold stage, a surprising result at the time. Since then Dimlington has been formalised as the type-section of the glaciation that occurred late in the last cold stage in the UK—the Late Devensian Dimlington Stadial (Rose, 1985). Catt (2007) has reviewed the importance of the site and research continues to the present day. The stratigraphic succession is given in Table 6.

The Basement Till, seen low in the cliff (Fig. 96), is the oldest till in Holderness (see also Itinerary 14, Loc. 3, Sewerby Buried Cliff). Where visible the till is matrix rich and of a very dark grey to olive grey colour. Whilst only a few metres of the till can usually be seen at the cliff foot, Catt and Digby (1988) reported about 30 m present in boreholes from the nearby gas terminal site, reaching almost to Chalk bedrock. Whilst it is generally considered to date to the penultimate glaciation ('Wolstonian') its elevation above modern sea level at Dimlington, together with amino acid analyses on included shell fragments in exposed sections has suggested a younger date (see Catt, 2007 for discussion), at least for the visible till. The upper levels of the till have been slightly weathered and sheared and folded suggesting that the shell fragments may have been introduced by the

157

Figure 96. Tills at Dimlington High Land [TA 392 215]. Olive grey Basement Till by the spade is overlain by dark greyish brown Skipsea Till (concealed by the slip plane) which is in turn overlain by dark brown Withernsea Till. Total section height about 24 m, spade 1.08 m.

overriding Skipsea Till ice sheet during the Late Devensian (Catt, 2007). Deformation structures and clast fabrics in the till, together with its content of erratics indicate the ice sheet flowed into the area from the north-east. Large masses (up to 100 m across) of grey, shelly, marine clay are incorporated as erratic rafts within the Basement Till and may occasionally be seen on the foreshore when storms have stripped the beach sand. Similar material is known from the Basement Till at Bridlington (though rarely seen). These rafts are termed the 'Bridlington Crag'. Amino acid (and palynological) analyses from the Bridlington site have yielded ratios from shells suggesting they date from the Early Pleistocene (Catt, 2007). The Bridlington Crag at Dimlington contains a range of erratics, including Larvikite and Rhomb Porphyry from the Oslo graben in Norway, presumably rafted into the marine clay by icebergs.

Overlying the Basement Till along the section are basins filled with a coarsening upward succession of silts and sands—the Dimlington Silts. These basins are of a glacitectonic origin and disrupt what was a more laterally extensive body of lacustrine sediment. They contain fragments of the moss *Pohlia wahlenbergi* var. *glacialis* together with an impoverished arctic beetle fauna and boreal freshwater ostracods. Collectively the fauna and flora indicate a shallow lake with little vegetation and surrounding bare ground. The moss remains have given C14 dates of ~18,240 and 18,500 years (Penny et al., 1969). Recent OSL dating of sands within the basins has given ages of 20.5–21.2 ka, comparable with the earlier C14 dates when calibrated (Bateman et al., 2015). These dates provide further evidence that the Basement Till is at least older than ~20 ka and that the Skipsea/Withernsea Till succession is younger than ~20 ka.

The Dimlington Silts were tectonised and disrupted by the advance of the North Sea Lobe of the last ice sheet after about 20 ka. The dark greyish-brown Skipsea Till (from 6–9 m thick: Fig. 96) overlies the silt/sand basins. In some sections younger silts and sands can be seen lying between the Skipsea and Withernsea Tills (Fig. 97). These sediments contain a sparse ichnofauna and delicately preserved adult and larval *Diptera* (flies). They accumulated in a sub-aerially exposed glacilacustrine environment and date to 17.1 ka (Bateman *et al*., 2015) and indicate that the ice sheet withdrew some distance to the east prior to re-advancing and depositing the overlying Withernsea Till (Fig. 97). Catt (2007) reported that the Withernsea Till unit is up to 24 m thick at Dimlington. Recent research by Evans and Thomson (2010) suggests that this great thickness is a product of glacitectonic stacking of till beds within an ice marginal environment, constructing the complex morainic topography of the area. A calibrated C14 date of 15.6 ka from basal lacustrine sediment overlying the Withernsea Till at a nearby site, The Bog, Roos, demonstrates the Skipsea/Withernsea tills complex was deposited by the ice sheet over a short, but dynamic, period between ~20 and ~15 ka in the Late Devensian before it finally withdrew from the area.

Figure 97. Bedded sands above Skipsea Till, Dimlington High Land [TA 391 217]. The sands are overlain by Withernsea Till.

4 REFERENCES

Agar, R. 1960. Post-glacial erosion of the north Yorkshire coast from the Tees estuary to the Humber. *Proceedings of the Yorkshire Geological Society*, **32**, 409–428.

Alexander, J. 1986. Idealised flow models to predict alluvial sandstone body distribution in the Middle Jurassic Yorkshire Basin. *Marine and Petroleum Geology*, **3**, 298–305.

Alexander, J. 1992. Nature and origin of a laterally extensive alluvial sandstone body in the Middle Jurassic Scalby Formation. *Journal of the Geological Society, London*, **149**, 431–441.

Allison, K.J. 1976. *The East Riding of Yorkshire Landscape*. Hodder & Stoughton, 272 pp.

Bailey, H.W., Gale, A.S., Mortimore, R.N., Swiecicki, A. & Wood, C.J. 1983. The Coniacian-Maastrichtian Stages in the United Kingdom, with particular reference to southern England. *Newsletters on Stratigraphy*, **12**, 19–42.

Bairstow, L.F. 1969. Lower Lias. In: Hemingway, J.E., Wright, J.K. & Torrens, H.S. (eds) *International Field Symposium on the British Jurassic. Excursion Guide 3, N.E. Yorkshire*. University of Keele, 47 pp.

Bate, R.H. 1959. The Yons Nab Beds of the Middle Jurassic of the Yorkshire Coast. *Proceedings of the Yorkshire Geological Society*, **32**, 153–164.

Bateman, M.D., Evans, D.J.A., Buckland, P.C., Connell, E.R., Friend, R.J., Hartmann, D., Moxon, H., Fairburn, W.A., Panagiotakopulu, E. and Ashurst, R.A. 2015. Last glacial dynamics of the Vale of York and North Sea lobes of the British and Irish Ice Sheet. *Proceedings of the Geologists' Association*, **126**, 712–730.

Berridge, N.G. & Pattison, J. 1994. Geology of the country around Grimsby and Patrington. *Memoir of the British Geological Survey*, Sheets 90, 91, 81 and 82 (England and Wales), xii +96 pp.

Bisat, W.S. 1962. Itinerary VIII. Hornsea to Mappleton. In: Bisat, W.S., Penny, L.F. & Neale, J.W. *Geology around the University Towns: Hull*. Geologists' Association Guides, No. **11**. Benham and Company Limited, Colchester, 34 pp.

Bisat, W.S., Penny, L.F. & Neale, J.W., 1962. *Geology around the University Towns: Hull*. Geologists' Association Guides, No. **11**. Benham and Company Limited, Colchester, 34 pp.

Black, M. 1928. 'Washouts' in the Estuarine Series of Yorkshire. *Geological Magazine*, **65**, 301–307.

Black, M. 1929. Drifted plant beds of the Upper Estuarine Series of Yorkshire. *Quarterly Journal of the Geological Society, London*, **85**, 389–437.

Black, M., Hemingway, J.E. & Wilson, V. 1934. Summer field meeting in N.E. Yorkshire: report by the directors. *Proceedings of the Geologists' Association*, **45**, 291–306.

Boston, C.M., Evans, D.J.A. and Ó Cofaigh, C. 2010. Styles of till deposition at the margin of the Last Glacial Maximum North Sea lobe of the British-Irish Ice Sheet: an assessment based on geochemical properties of glacigenic deposits in eastern England, *Quaternary Science Reviews*, **29**, 3184–3211.

Bott, M.H.P., Robinson, J. & Kohnstamm, M.A. 1978. Granite beneath Market Weighton, east Yorkshire. *Journal of the Geological Society, London*, **135**, 535–543.

Bray, R.J., Green, P.F. & Duddy, L.R. 1992. Thermal history reconstruction using apatite fission track analysis and vitrinite reflectance: a case study from the U.K. East Midlands and Southern North Sea. *Geological Society Special Publication*, **67**, 3–25.

British Geological Survey 1998a. *Whitby and Scalby*. England and Wales Sheet 35 & 44. 1:50 000 Provisional Series.

British Geological Survey 1998b. *Scarborough*. England and Wales Sheet 54. 1:50 000 Provisional Series.

4 References

Buckman, S.S. 1915. A palaeontological classification of the Jurassic rocks of the Whitby district, with a zonal table of Liassic ammonites. In: Fox-Strangways, C. & Barrow, G. *The Geology of the country between Whitby and Scarborough*. Memoirs of the Geological Survey. England and Wales (2nd edition), 59–102.

Busfield, M.E., Lee, J.R., Riding, J.B., Zalasiewicz, J. & Lee, S.V. 2015. Pleistocene till provenance in east Yorkshire; reconstructing ice flow of the British North Sea Lobe. *Proceedings of the Geologists' Association*, **126**, 86–99.

Caswell, B.A., Coe, A.L. & Cohen, A.S. 2009. New range data for marine invertebrate species across the early Toarcian (Early Jurassic) mass extinction. *Journal of the Geological Society, London*, **166**, 839–872.

Callomon, J.H. & Wright, J.K. 1989. Cardioceratid and Kosmoceratid ammonites from the Callovian of Yorkshire. *Palaeontology*, **32**, 799–836.

Catt, J.A. 2007. The Pleistocene glaciations of eastern Yorkshire: a review. *Proceedings of the Yorkshire Geological Society*, **56**, 177–207.

Catt, J.A. & Digby, P.G.N. 1988. Boreholes in the Wolstonian Basement Till at Easington, Holderness, July 1985. *Proceedings of the Yorkshire Geological Society*, **47**, 21–27.

Catt, J.A. & Madgett, P.A. 1981. The Work of W.S.Bisat F.R.S. on the Yorkshire Coast. In: Neale, J. & Flenley, J. (eds) *The Quaternary in Britain. Essays, reviews and original work on the Quaternary published in honour of Lewis Penny on his retirement*. Pergamon Press, Oxford. 119–136.

Catt, J.A. & Penny, L.F. 1966. The Pleistocene deposits of Holderness, East Yorkshire. *Proceedings of the Yorkshire Geological Society*, **35**, 375–420.

Catt, J.A., Weir, A.H. & Madgett, P.A. 1974. The loess of eastern Yorkshire and Lincolnshire. *Proceedings of the Yorkshire Geological Society*, **40**, 23–39.

Chapman, S., 2002. *Grosmont and its Mines*. P. Tuffs, Guisborough, 68pp.

Clark, J. 1954. *Excavations at Star Carr: An Early Mesolithic Site at Seamer near Scarborough, Yorkshire*. Cambridge University Press, Cambridge.

Coe, A.L., 1996. A comparison of the Oxfordian successions of Dorset, Oxfordshire and Yorkshire. In: Taylor, P.D. (ed.). *Field geology of the British Jurassic*. Geological Society, London. 151–172.

Cohen, A.S., Coe, A.L. & Kemp, D.B. 2007. The Late Paleocene-Early Eocene and Toarcian (Early Jurassic) carbon isotope excursions: a comparison of their time scales, associated environmental changes, causes and consequences. *Journal of the Geological Society, London*, **164**, 1093–1108.

Cox, B.M. & Richardson, G. 1982. The ammonite zonation of Upper Oxfordian mudstones in the Vale of Pickering, Yorkshire. *Proceedings of the Yorkshire Geological Society*, **44**, 53–58.

Danise, S., Twitchett, R.J. & Little, C.T.S. 2015. Environmental controls on Jurassic marine ecosystems during global warming. *Geology*, **43**, 263–266.

Dakyns, J.R. 1880. Glacial deposits north of Bridlington. *Proceedings of the Yorkshire Geological & Polytechnic Society*, **7**, 246–252.

Dean, W.T. 1954. Notes on part of the Upper Lias succession at Blea Wyke, Yorkshire. *Proceedings of the Yorkshire Geological Society*, **29**, 161–179.

Delair, J.B. & Sargeant, W.A.S. 1985. History and bibliography of the study of fossil vertebrate footprints in the British Isles: Supplement 1973-83. *Palaeogeography, Palaeoclimatology, Palaeoecology*, **19**, 123–160.

Donato, J.A. 1993. A buried granite batholith and the origin of the Sole Pit Basin, UK Southern North Sea. *Journal of the Geological Society, London*, **150**, 255–258.

Dove, D., Evans, D.J.A., Lee, J.R., Roberts, D.H., Tappin, D.R., Mellett, C.L., Long, D. & Callard, S.L. 2017. Phased occupation and retreat of the last British-Irish Ice Sheet

in the southern North Sea; geomorphic and seismostratigraphic evidence of a dynamic ice lobe. *Quaternary Science Reviews*, **163**, 114–134.
Edwards, C.A. 1981. The Tills of Filey Bay. In: J. Neale & J. Flenley (eds), *The Quaternary in Britain. Essays, reviews and original work in honour of Lewis Penny on his retirement*. Pergamon Press, Oxford, 108–118.
Evans, D.J.A., Bateman, M.D., Roberts, D.H., Medialdea, A., Hayes, L., Duller, G.A.T., Fabel, D. & Clark, C.D. 2017. Glacial Lake Pickering: stratigraphy and chronology of a proglacial lake dammed by the North Sea Lobe of the British-Irish Ice Sheet. *Journal of Quaternary Science*, **32**, 295–310.
Evans, D.J.A., Owen, L.A. & Roberts, D. 1995. Stratigraphy and sedimentology of Devensian (Dimlington Stadial) glacial deposits, east Yorkshire, England. *Journal of Quaternary Science*, **10**, 241–265.
Evans, D.J.A. & Thomson, S.A. 2010. Glacial sediments and landforms of Holderness, eastern England: A glacial depositional model for the North Sea Lobe of the British-Irish Ice Sheet. *Earth-Science Reviews*, **101**, 147–189.
Fish, P.R., Moore, R. & Carey, J.M. 2006. Landslide geomorphology of Cayton Bay, North Yorkshire, UK. *Proceedings of the Yorkshire Geological Society*, **56**, 5–14.
Fletcher, B.N. 1969. A lithological subdivision of the Speeton Clay C Beds (Hauterivian), East Yorkshire. *Proceedings of the Yorkshire Geological Society*, **37**, 323–327.
Fox-Strangways, C. 1892. The Jurassic rocks of Great Britain, vol. 1. Yorkshire. *Memoirs of the Geological Survey of the United Kingdom*. ix + 551 pp.
Fox-Strangways, C. & Barrow, G. 1915. The Geology of the country between Whitby and Scarborough. *Memoirs of the Geological Survey. England and Wales*. (2nd edition). iv + 144 pp.
Gilbertson, D.D. (ed.) 1984. *Late Quaternary Environments and Man in Holderness*. BAR British Series **134**. xviii + 243pp.
Godfrey, A. & Lassey, P.J. 1974. *Shipwrecks of the Yorkshire Coast*. Dalesman Books, 168 pp.
Goldring, D. 2001. *Along the Scar. A Guide to the Mining Geology and Industrial Archaeology of the North Yorkshire Coast*. P. Tuffs, Guisborough, 145 pp.
Goldring, D. 2006. *Along the Esk. A Guide to the Mining Geology and Industrial Archaeology of the Esk Valley*. P. Tuffs, Guisborough, 168 pp.
Gowland, S. & Riding, J.B. 1991. Stratigraphy, sedimentology and palaeontology of the Scarborough Formation (Middle Jurassic) of Hundale Point, North Yorkshire. *Proceedings of the Yorkshire Geological Society*, **48**, 375–392.
Greensmith, J.T., Rawson, P.F. & Shalaby, S.E. 1980. An association of minor fining-upwards cycles and aligned gutter marks in the Middle Lias (Lower Jurassic) of the Yorkshire Coast. *Proceedings of the Yorkshire Geological Society*, **42**, 525–538.
Gregory, K.J. 1962. The deglaciation of eastern Eskdale, Yorkshire. *Proceedings of the Yorkshire Geological Society*, **33**, 363–380.
Gregory, K.J. 1965. Proglacial Lake Eskdale after 60 years. *Transactions of the Institute of British Geographers*, **36**, 149–162.
Hargreaves, J.A. 1914. Fossil footprints near Scarborough. *The Naturalist*, **1914**, 9–95.
Harris, T.M. 1953. The Geology of the Yorkshire flora. *Proceedings of the Yorkshire Geological Society*, **29**, 63–71.
Harris, T.M. 1961. *The Yorkshire Jurassic flora, Volume 2, Caytoniales, Cycadales and Pteridosperms*. British Museum (Natural History), London, 191 pp 7 pls.
Hart, M.B. 2018. The 'Black Band': local expression of a global event. *Proceedings of the Yorkshire Geological Society*, **62**, (doi.org/10.1144/pygs2017-007).
Hayes, R.H. & Rutter, J.G. 1974. Rosedale mines and railway. *Scarborough Archaeological and Historical Society, Research Report* no. **9**, 32 pp.

Hemingway, J.E. 1958. The Geology of the Whitby area. In: Daysh, G.H.J. (ed.) *A survey of Whitby and the surrounding area*. The Shakespeare Head, Eton. 1–47.
Hemingway, J.E. 1968. Egton Bridge and Goathland. In: Hemingway, J.E., Wilson, V. & Wright, C.W. *Geology of the Yorkshire Coast*. Geologists' Association Guide No. **34**, 28–31.
Hemingway, J.E. 1974a. Jurassic. In: Rayner, D.H. & Hemingway, J.E. (eds). *The Geology and Mineral Resources of Yorkshire*. Yorkshire Geological Society, Leeds, 161–223.
Hemingway, J.E. 1974b. Ironstone. In Rayner, D.H. & Hemingway, J.E. (eds). *The Geology and Mineral Resources of Yorkshire*. Yorkshire Geological Society, Leeds, 329–335.
Hemingway, J.E. & Riddler, G.P. 1982. Basin inversion in North Yorkshire. *Transactions of the Institution of Mining and Metallurgy (section B)*, **91**, B175–B186.
Hemingway, J.E., Wilson, V & Wright, C.W., 1963. *Geology of the Yorkshire Coast*. Geologists' Association Guide No. **34**, 34 pp.
Hemingway, J.E., Wilson, V & Wright, C.W., 1968. *Geology of the Yorkshire Coast*. Geologists' Association Guide No. **34**, 47 pp.
Herbin, J.-P., Muller, C., Geyssant, J.R., Méliéres, F. & Penn, I.E. 1991. Hétérogénéité quantitative et qualitative de la matière organique dans les argiles du Kimmeridgien du Val de Pickering (Yorkshire, UK). *Revue de l'Institut Français du Pétrole*, **46**, 675–712.
Hesselbo, S.P. & Jenkyns, H.C., 1996. A comparison of the Hettangian to Bajocian successions of Dorset and Yorkshire. In: Taylor, P.D. (ed.). *Field Geology of the British Jurassic*. Geological Society, London, 105–150.
Hesselbo, S.P., Morgans-Bell, H.S., McElwain, J.C., Rees, P. McA., Robinson, S.A. & Ross, C.E., 2003. Carbon-cycle perturbation in the Middle Jurassic and accompanying changes in the terrestrial environment. *Journal of Geology*, **111**, 259–276.
Hildreth, P.N. 2018. The distribution and form of flint, with particular reference to the Chalk Group (Upper Cretaceous) of the Northern Province, UK. *Proceedings of the Yorkshire Geological Society*, **62**, (doi.org/10.1144/pygs2017-383).
Howard, A.S. 1985. Lithostratigraphy of the Staithes Sandstone and Cleveland Ironstone Formations (Lower Jurassic) of north-east Yorkshire. *Proceedings of the Yorkshire Geological Society*, **45**, 261–275.
Howarth, M.K. 1955. Domerian of the Yorkshire Coast. *Proceedings of the Yorkshire Geological Society*, **30**, 147–175.
Howarth, M.K. 1962. The Jet Rock Series and the Alum Shale Series of the Yorkshire Coast. *Proceedings of the Yorkshire Geological Society*, **33**, 381–422.
Howarth, M.K. 1973. The stratigraphy and ammonite fauna of the Upper Liassic Grey Shales of the Yorkshire Coast. *Bulletin of the British Museum (Natural History)*, **24**, 235–277.
Howarth, M.K. 1992. The ammonite family Hildoceratidae in the Lower Jurassic of Britain. Part 1. *Monograph of the Palaeontographical Society*, 1–106.
Howarth, M.K. 2002. The Lower Lias of Robin Hood's Bay, Yorkshire, and the work of Leslie Bairstow. *Bulletin of the Natural History Museum London (Geology)*, **58 (2)**, 81–152.
Hudleston, W.H. 1878. The Yorkshire Oolites, Pt 2, the Middle Oolites. Section 2, the Coralline Oolites, Coral Rag and Supra-Coralline Beds. *Proceedings of the Geologists' Association*, **5**, 407–494.
Hughes, F., Harrison, D., Haarhoff, M., Howlett, P., Pearson, A., Ware, D., Taylor, C., Emms, G. & Mortimer, A. 2016. The unconventional Carboniferous reservoirs of the Greater Kirby Misperton gas field and their potential: North Yorkshire's sleeping giant. In: Bowman, D. & Levell, B. (eds) *Petroleum Geology of NW Europe: 50 years*

of Learning—Proceedings of the 8th Petroleum Geology Conference. Geological Society of London, 611–625.
Ielpi, A. & Ghinassi, M. 2014. Planform architecture, stratigraphic signature and morphodynamics of an exhumed Jurassic meander plain (Scalby Formation, Yorkshire, UK). *Sedimentology*, **61**, 1923–1960.
Ivens, C.R. & Watson, G.G. 1994. *Records of dinosaur footprints on the North East Yorkshire Coast*. Roseberry Publications, Middlesborough, 20 pp.
Jackson, J.W. 1911. A new species of *Unio* from the Yorkshire Estuarine Series. *The Naturalist*, **1911**, 211–214.
Jeans, C.V., Long, D., Hu, X.-F. & Mortimore, R. 2014. Regional hardening of Upper Cretaceous Chalk in eastern England, UK: trace element and stable isotope patterns in the Upper Cenomanian and Turonian Chalk and their significance. *Acta Geologica Polonica*, **64**, 419–455.
Johnson, R.M. & Fish, P.R. 2012. Reactivation of the coastal landslide system at Cayton Bay, North Yorkshire, UK. *Proceedings of the Yorkshire Geological Society*, **59**, 77–89.
Kantorowicz, J.D. 1990. Lateral and vertical variation in pedogenesis and other early diagenetic phenomena, Middle Jurassic Ravenscar group, Yorkshire. *Proceedings of the Yorkshire Geological Society*, **48**, 61–74.
Kaye, P. 1964. Observations on the Speeton Clay (Lower Cretaceous). *Geological Magazine*, **101**, 340–356.
Kelly, S.R.A., Gregory, J., Braham, W., Strogan, D.P. & Whitham, A.G. 2015. Towards an integrated Jurassic biostratigraphy for East Greenland. *Volumina Jurassic*, **13**, 43–64.
Kemper, E., Rawson, P. F. & Thieuloy, J.-P. 1981. Ammonites of Tethyan ancestry in the early Lower Cretaceous of north-west Europe. *Palaeontology*, **24**, 251–311.
Kendall, P.F. 1902. A system of glacier lakes in the Cleveland Hills. *Quarterly Journal of the Geological Society, London*, **58**, 471–571.
Kent, P.E. 1980a. Subsidence and uplift in East Yorkshire and Lincolnshire: a double inversion. *Proceedings of the Yorkshire Geological Society*, **42**, 505–524.
Kent. P.E. 1980b. *Eastern England from the Tees to the Wash*. British Regional Geology, HMSO, vii + 155 pp.
Kirby, G.A. & Swallow, P.W. 1987. Tectonism and sedimentation in the Flamborough Head region of north-east England. *Proceedings of the Yorkshire Geological Society*, **46**, 301–309.
Knox, R.W.O'B. 1973. The Eller Beck Bed (Bajocian) of the Ravenscar Group of north-east Yorkshire. *Geological Magazine*, **110**, 511–534.
Knox, R.W.O'B. 1984. Lithostratigraphy and depositional history of the late Toarcian sequence at Ravenscar, Yorkshire. *Proceedings of the Yorkshire Geological Society*, **45**, 99–108.
Knox, R.W.O'B. 1991. Ryazanian to Barremian mineral stratigraphy of the Speeton Clay in the southern North Sea Basin. *Proceedings of the Yorkshire Geological Society*, **48**, 255–264.
Knox, R.W.O'B., Howard, A.S., Powell, J.H. & Van Buchem, F. 1991. *Lower and Middle Jurassic sediments of the Cleveland Basin, N.E. England: shallow marine and paralic facies seen in their sequence stratigraphic context*. 13th International Sedimentological Congress, Field Guide No. **5**.
Konijnenburg-van Cittert, J.H.A. van & Morgans, H.S. 1999. *The Jurassic Flora of Yorkshire*. Palaeontological Association field guides to fossils, **8**, 134 pp.
Lamplugh, G.W. 1889. On the subdivisions of the Speeton Clay. *Quarterly Journal of the Geological Society of London*, **45**, 575–618.

Lamplugh, G.W. 1890. On a new locality for the Arctic Fauna of the "Basement" Boulder Clay in Yorkshire. *Geological Magazine*, **7**, 61–70.
Lamplugh, G.W. 1891a. On the Boulders and Glaciated Rock Surfaces of the Yorkshire Coast. *Report of the British Association for the advancement of Science (for 1890)*, 797–798.
Lamplugh, G.W. 1891b. On the drifts of Flamborough Head. *Quarterly Journal of the Geological Society of London*, **47**, 384–431.
Lamplugh, G.W. 1896. Notes on the White Chalk of Yorkshire. Part III. The Geology of Flamborough Head, with notes on the Yorkshire Wolds. *Proceedings of the Yorkshire Geological Society*, **13**, 171–191.
Livera, S.E. & Leeder, M.R. 1981. The Middle Jurassic Ravenscar Group ("Deltaic Series") of Yorkshire: recent sedimentological studies as demonstrated during a field meeting, 2-3 May 1980. *Proceedings of the Geologists' Association*, **92**, 241–250.
McElwain, J.C., Wade-Murphy, J. & Hesselbo, S.P. 2005. Changes in carbon dioxide during an oceanic anoxic event linked to intrusion into Gondwana coals. *Nature*, **435**, 479–482.
Meister, C., Aberhan, M., Blau, J., Dommergues, J.-L., Feist-Burkhardt, S., Hailwood, E.A., Hart, M., Hesselbo, S.P., Hounslow, M.W., Hylton, M., Morton, N., Page, K. & Price, G.D. 2006. The Global Stratotype Section and Point for the base of the Pliensbachian Stage (Lower Jurassic), Wine Haven, Yorkshire, UK. *Episodes*, **29**, 93–106.
Middlemiss, F.A. 1976. Lower Cretaceous Terebratulinida of Northern England and Germany and their geological background. *Geologisches Jährbuch*, **A30**, 21–104.
Milner, N., Conneller, C., Taylor, B. & Schadla-Hall, R.T. 2012. *The Story of Star Carr*. Council for British Archaeology, 20 pp.
Milsom, J. & Rawson, P.F. 1989. The Peak Trough - a major control on the geology of the North Yorkshire coast. *Geological Magazine*, **126**, 699–705.
Mitchell, S.F. 1994. New data on the biostratigraphy of the Flamborough Chalk Formation (Santonian, Upper Cretaceous) between South Landing and Danes Dyke, North Yorkshire. *Proceedings of the Yorkshire Geological Society*, **50**, 113–118.
Mitchell, S.F. 1995. Lithostratigraphy and biostratigraphy of the Hunstanton Formation (Red Chalk, Cretaceous) succession at Speeton, North Yorkshire, England. *Proceedings of the Yorkshire Geological Society*, **50**, 285–303.
Mortimore, R.N., Wood, C.J. & Gallois, R.W. 2001. *British Upper Cretaceous Stratigraphy*. Geological Conservation Review Series, no. **23**, Joint Nature Conservation Committee, Peterborough, 558 pp.
Nami, M. 1976. An exhumed Jurassic meander belt from Yorkshire. *Geological Magazine*, **113**, 47–52.
Nami, M. & Leeder, M.R. 1978. Changing channel morphologies and magnitude in the Scalby Formation (Middle Jurassic) of Yorkshire (England). In: Miall, A.D. (ed.) *Fluvial sedimentation*. Canadian Society of Petroleum Geologists, Memoir **5**, 431–440.
Neale, J.W. 1960. The subdivision of the Upper D Beds of the Speeton Clay of Speeton, East Yorkshire. *Geological Magazine*, **97**, 353–362.
Neale, J.W. 1962. Ammonoidea from the lower D Beds (Berriasian) of the Speeton Clay. *Palaeontology*, **5**, 272–296.
Ogg, J.G. & Hinov, L.A. 2012. Jurassic (Chapter 26). In: Gradstein, F.M., Ogg, J.G., Schitz, M. & Gradstein, G. (eds) *The Geologic Timescale*, Vol. 2. Elsevier BV, Amsterdam, 731–791.
Osborne, R. 1998. *The Floating Egg*. Jonathan Cape, London, xii + 372 pp.
Owen, J.S. 1979. The Cleveland Ironstone Mining Industry. In: *Cleveland Iron and Steel*. British Steel Corporation, 9–48. (Reprinted as *Cleveland Ironstone Mining* by C. Books, Redcar, 1986, and by Tom Leonard Mining Museum 1995.)

Owen, J.S. 1985. *Staithes and Port Mulgrave Ironstone*. Cleveland Industrial Archaeology Society. Research Report no. **4**, 41 pp.

Owen, J.S. 1988. *The Ironworks at Runswick*. Cleveland Industrial Archaeological Society Research Report no. **5**, 29 pp.

Page, K.N. 1989. A stratigraphic revision for the English Lower Callovian. *Proceedings of the Geologists' Association*, **100**, 363–382.

Palfry, J. & Smith, P.L. 2000. Synchrony between Early Jurassic extinction, oceanic oxygen event, and the Karoo-Ferrar flood basalt volcanism. *Geology*, **28**, 747–750.

Palmer, A.P., Matthews, I.P., Candy, I., Blockley, S.P.E., MacLeod, A., Darvill, C.M., Milner, N., Conneller, C. & Taylor, B. 2015. The evolution of Palaeolake Flixton and the environmental context of Star Carr, NE. Yorkshire: stratigraphy and sedimentology of the Last Glacial-Interglacial Transition (LGIT) lacustrine sequences. *Proceedings of the Geologists' Association*, **126**, 50–59.

Parsons, C.F. 1977. A stratigraphical revision of the Scarborough Formation. *Proceedings of the Yorkshire Geological Society*, **41**, 203–222.

Penny, L.F., Coope, G.R. & Catt, J.A. 1969. Age and Insect Fauna of the Dimlington Silts, East Yorkshire. *Nature*, **224**, no. 5214, 65–67.

Penny, L.F. & Rawson, P.F. 1969. Field meeting in East Yorkshire and North Lincolnshire. *Proceedings of the Geologists' Association*, **80**, 193–218.

Phillips, J. 1829. *Illustrations of the Geology of Yorkshire; or, a description of the strata and organic remains of the Yorkshire Coast*. London, xvi + 192 pp. (2nd edit. 1835, 3rd edit. 1875)

Powell, J.H., 1984. Lithostratigraphic nomenclature of the Lias Group in the Yorkshire Basin. *Proceedings of the Yorkshire Geological Society*, **45**, 51–57.

Powell, J.H. 2010. Jurassic sedimentation in the Cleveland Basin: a review. *Proceedings of the Yorkshire Geological Society*, **58**, 21–72.

Powell, J. H., Ford, J.R. & Riding, J.B. 2016. Diamicton from the Vale of York and Tabular Hills, north-east Yorkshire: evidence for a Middle Pleistocene (MIS8) glaciation? *Proceedings of the Geologists' Association*, **127**, 575–594.

Powell, J.H., Rawson, P.F., Riding, J.B. & Ford, J.R. 2018. Sedimentology and stratigraphy of the Kellaways Sand Member (Lower Callovian), Burythorpe, North Yorkshire, UK. *Proceedings of the Yorkshire Geological Society*, **62**, 36–49.

Pybus, D. & Rushton, J. 1991. Alum and the Yorkshire coast. In: Lewis, D. B. (ed.) *The Yorkshire Coast*. Normandy Press, 46–59.

Pye, K. & Krinsley, D.H. 1986. Microfabric, mineralogy and early diagenetic history of the Whitby Mudstone Formation (Toarcian), Cleveland Basin, U.K. *Geological Magazine*, **123**, 191–203.

Rastall, R.H. & Hemingway, J.E. 1940. The Yorkshire Dogger, 1. The coastal region. *Geological Magazine*, **77**, 177–197.

Ravenne, Ch. 2002. Stratigraphy and Oil: A Review. Part 2. Characterization of Reservoirs and Sequence Stratigraphy: Quantification and Modeling. *Oil & Gas Science and Technology – Rev. IFP*, **57** **(4)**, 311–340.

Rawson, P. F. 1971. Lower Cretaceous ammonites from north-east England: the Hauterivian genus *Simbirskites*. *Bulletin of the British Museum (Natural History) Geology*, **20**, 25–86.

Rawson, P.F. 1975. Lower Cretaceous ammonites from north-east England: the Hauterivian heteromorph *Aegocrioceras*. *Bulletin of the British Museum (Natural History) Geology*, **26**, 129–159.

Rawson, P. F. 1995. The "Boreal" Early Cretaceous (Pre-Aptian) ammonite sequences of NW Europe and their correlation with the Western Mediterranean faunas. *Memorie Descrittive della Carta Geologica d'Italia*, **51**, 121–130.

Rawson, P.F. 2006. Cretaceous: sea levels peak as the North Atlantic opens. In: Brenchley, P.J. & Rawson, P.F. (eds) *The Geology of England and Wales* (2nd edition). Geological Society, London, 365–393.

Rawson, P.F. 2007. Global relationships of Argentine (Neuquén Basin) Early Cretaceous ammonite faunas. *Geological Journal*, **42**, 175–183.

Rawson, P.F. 2014. Hauterivian correlations. In: Reboulet, S. *et al.* Report on the 5th International Meeting of the IUGS Lower Cretaceous Ammonite Group, the "Kilian Group" (Ankara, Turkey, 31st August 2013). *Cretaceous Research*, **50**, 126–137.

Rawson, P.F., Greensmith, J.T. & Shalaby, S.E. 1983. Coarsening upwards cycles in the uppermost Staithes and Cleveland Ironstone Formations (Lower Jurassic) of the Yorkshire coast, England. *Proceedings of the Geologists' Association*, **94**, 91–93.

Rawson, P.F. & Mutterlose, J. 1983. Stratigraphy of the Lower B and basal Cement Beds (Barremian) of the Speeton Clay, Yorkshire, England. *Proceedings of the Geologists' Association*, **94**, 133–146.

Rawson, P.F. & Riley, L.A. 1982. Latest Jurassic - Early Cretaceous events and the "Late Cimmerian Unconformity" in North Sea area. *Bulletin of the American Association of Petroleum Geologists*, **66**, 2628–2648.

Rawson, P.F. & Wright, J.K. 1992. *Geology of the Yorkshire Coast*. Geologists' Association Guide No. **34**, 117 pp.

Rawson, P.F. & Wright, J.K. 1996. Jurassic of the Cleveland Basin, North Yorkshire. In: Taylor, P.D. (ed.). *Field Geology of the British Jurassic*. Geological Society, London, 173–208.

Rawson, P.F. & Wright, J.K. 2000. *Geology of the Yorkshire Coast*. Geologists' Association Guide No. **34**, 130 pp.

Riding, J.B. 1984. A palynological investigation of Toarcian to early Aalenian strata from the Blea Wyke area, Ravenscar, North Yorkshire. *Proceedings of the Yorkshire Geological Society*, **45**, 109–122.

Riding, J.B. & Wright, J.K. 1989. Palynostratigraphy of the Scalby Formation (Middle Jurassic) of the Cleveland basin, north-east Yorkshire. *Proceedings of the Yorkshire Geological Society*, **47**, 349–354.

Roberts, D.H., Evans, D.J.A., Lodwick, J. & Cox, N.J. 2013. The subglacial and ice-marginal signature of the North Sea Lobe of the British-Irish Ice Sheet during the Last Glacial Maximum at Upgang, North Yorkshire, UK. *Proceedings of the Geologists' Association*, **124**, 503–519.

Romano, M. & Whyte, M.A. 2003. Jurassic dinosaur tracks and trackways of the Cleveland Basin, Yorkshire: preservation, diversity and distribution. *Proceedings of the Yorkshire Geological Society*, **54**, 185–215.

Romano, M. & Whyte, M.A. 2013. A new record of the trace fossil *Selenichnites* from the Middle Jurassic Scalby Formation of the Cleveland Basin, Yorkshire. *Proceedings of the Yorkshire Geological Society*, **59**, 203–210.

Romano, M. & Whyte, M.A. 2015. Could stegosaurs swim? Suggestive evidence from the Middle Jurassic tracksite of the Cleveland Basin, Yorkshire, UK. *Proceedings of the Yorkshire Geological Society*, **60**, 227–233.

Romano, M., Whyte, M.A. & Manning, P.L. 1999. New sauropod dinosaur prints from the Saltwick Formation (Middle Jurassic) of the Cleveland Basin, Yorkshire. *Proceedings of the Yorkshire Geological Society*, **52**, 361–369.

Rosales, I., Quesada, S. & Robles, S. 2004. Palaeotemperature variations of Early Jurassic seawater recorded in geochemical trends of belemnites from the Basque-Cantabrian basin, northern Spain. *Palaeogeography, Palaeoclimatology, Palaeoecology*, **203**, 252–275.

4 References

Rose, J. 1985. The Dimlington Stadial/Dimlington Chronozone: a proposal for naming the main glacial episode of the Late Devensian in Britain. *Boreas*, **14**, 225–230.

Rowe, A.W. 1904. The zones of the White Chalk of the English coast. IV - Yorkshire. *Geological Magazine*, **99**, 273–278.

Sargeant, W.A.S. 1970. Fossil footprints from the Middle Trias of Nottinghamshire and the Middle Jurassic of Yorkshire. *Mercian Geologist*, **3**, 269–282.

Schlanger, S.O. & Jenkyns, H.C. 1976. Cretaceous oceanic anoxic events: Causes and consequences. *Geologie en Mijnbouw*, **55**, 179–184.

Sellwood, B.W. 1970. The relation of trace fossils to small-scale sedimentary cycles in the British Lias. *Geological Journal Special Issue*, **3**, 489–504.

Senior, J.R. 1994. The Lower Jurassic rocks between Staithes and Port Mulgrave. In: Scrutton, C. (ed.) *Yorkshire rocks and landscape. A field guide*. Yorkshire Geological Society, 224 pp.

Smith, W. 1829–30. Memoir of the Stratification of the Hackness Hills. Published in: Fox-Strangways, C. 1892. *The Jurassic rocks of Great Britain, vol. 1. Yorkshire*. Memoirs of the Geological Survey of the United Kingdom, 507 pp.

Starmer, I.C. 1995a. Deformation of the Upper Cretaceous Chalk at Selwicks Bay, Flamborough Head, Yorkshire: its significance in the structural evolution of north-east England and the North Sea Basin. *Proceedings of the Yorkshire Geological Society*, **50**, 213–228.

Starmer, I.C. 1995b. Contortions in the Chalk at Staple Nook, Flamborough Head. *Proceedings of the Yorkshire Geological Society*, **50**, 271–275.

Starmer, I.C. 2013. Folding and faulting in the Chalk at Dykes End, Bridlington Bay, East Yorkshire, resulting from reactivations of the Flamborough Head Fault Zone. *Proceedings of the Yorkshire Geological Society*, **59**, 195–201.

Stather, J.W. 1897. A glaciated surface at Filey. *Proceedings of the Yorkshire Geological Society*, **13**, 346–349.

Stone, P., McMillan, A.A., Floyd, J.D., Barnes, R.P., & Phillips, E.R. 2012. *British regional geology: South of Scotland*. Fourth edition. British Geological Survey, 247 pp.

Straw, A. & Clayton, K. 1979. *The geomorphology of the British Isles. Eastern and Central England*. Methuen & Co. Ltd., London. 247 pp.

Sumbler, M.G. 1996. *The stratigraphy of the Chalk Group in Yorkshire, Humberside and Lincolnshire*. British Geological Survey, Technical Report, WA/96/26C.

Svenson, H., Planke, S., Chevallier, L., Malthe-Sorenssen, A., Corfu, F. & Jamtveit, B. 2007. Hydrothermal venting of greenhouse gasses triggered Early Jurassic global warming. *Earth and Planetary Sciences Letters*, **256**, 554–566.

Thomas, C. 2013–2015. The Moorland Collieries of North Yorkshire. Cleveland Industrial Heritage, **32**, 2–13; **33**, 2–13; **34**, 2–13; **35**, 2–13; **36**, 2–11.

Tucker, M.E. 1991. Sequence stratigraphy of carbonate – evaporite basins: models and applications to the Upper Permian (Zechstein) of northeast England and adjoining North Sea. *Journal of the Geological Society, London*, **148**, 1019–1036.

Van Buchem, F.S.P. & McCave, I.N. 1989. Cyclic sedimentation patterns in Lower Lias mudstones of Yorkshire (Great Britain). *Terra Nova*, **1**, 461–467.

Van Buchem, F.S.P., Melnyk, D.H. & McCave, I.N. 1992. Chemical cyclicity and correlation of Lower Lias Mudstones using gamma ray logs, Yorkshire, UK. *Journal of the Geological Society, London*, **149**, 991–1002.

Van Buchem, F.S.P., McCave, I.N. & Weedon, G.P. 1994. Orbitally induced small scale cyclicity in a siliciclastic epicontinental setting (Cleveland Basin, Lower Lias, Yorkshire, UK). In: De Boer, P.L. & Smith, D.G. (eds) *Orbital forcing and cyclic sedimentary sequences*. Special Publications of the International Association of Sedimentologists, **19**, 345–366.

Van Buchem, F.S.P. & Knox, R.W.O'B. 1998. Lower and Middle Jurassic depositional sequences of Yorkshire (UK). In: De Graciansky, P.C., Hardenbol, J., Jacquin, T. & Vail, P.R. (eds) *Mesozoic-Cenozoic Sequence Stratigraphy of European Basins*. SEPM, Special Publications, **60**, 545–559.
Versey, H.C. 1939. The Tertiary History of East Yorkshire. *Proceedings of the Yorkshire Geological Society*, **23**, 302–314.
Whitham, F. 1991. The stratigraphy of the Upper Cretaceous Ferriby, Welton and Burnham Formations north of the Humber, north-east England. *Proceedings of the Yorkshire Geological Society*, **48**, 227–254.
Whitham, F. 1993. The stratigraphy of the Upper Cretaceous Flamborough Chalk Formation north of the Humber, north-east England. *Proceedings of the Yorkshire Geological Society*, **49**, 235–258.
Whyte, M.A. & Romano, M. 1993. Footprints of a sauropod dinosaur from the Middle Jurassic of Yorkshire. *Proceedings of the Geologists' Association*, **104**, 195–199.
Whyte, M.A. & Romano, M. 2001. Probable stegosaurian dinosaur trails from the Saltwick Formation (Middle Jurassic) of Yorkshire, England. *Proceedings of the Geologists' Association*, **112**, 45–54.
Whyte, M.A., Romano, M., Hudson, J.G. & Watts, W. 2006. Discovery of the largest theropod dinosaur track known from the Middle Jurassic of Yorkshire. *Proceedings of the Yorkshire Geological Society*, **56**, 77–80.
Whyte, M.A., Romano, M. & Watts, W. 2010. Yorkshire dinosaurs: a history in two parts. In: Moody, R.T.J., Buffetaut, E., Naish, D. & Martill, D.M. (eds) *Dinosaurs and other Extinct Saurians: a Historical Perspective*. Geological Society of London. Special Publications, **343**, 189–207.
Wilson, V. 1949. The lower Corallian rocks of the Yorkshire coast and Hackness Hills. *Proceedings of the Geologists' Association*, **60**, 325–271.
Wood, C.J. & Smith, D. 1978. Lithostratigraphic nomenclature of the Chalk in North Yorkshire, Humberside and Lincolnshire. *Proceedings of the Yorkshire Geological Society*, **42**, 263–287.
Wray, D.S. & Wood, C.J. 1998. Distinction between detrital and volcanogenic clay-rich beds in Turonian-Coniacian chalks of eastern England. *Proceedings of the Yorkshire Geological Society*, **52**, 95–105.
Wright, J.K. 1968a. The stratigraphy of the Callovian rocks between Newtondale and the Scarborough coast, Yorkshire. *Proceedings of the Geologists' Association*, **79**, 363–399.
Wright, J.K. 1968b. The Callovian succession at Peckondale Hill, Malton, Yorkshire. *Proceedings of the Yorkshire Geological Society*, **37**, 93–97.
Wright, J. K. 1972. The stratigraphy of the Yorkshire Corallian. *Proceedings of the Yorkshire Geological Society*, **39**, 225–266.
Wright, J.K. 1977. The Cornbrash Formation (Callovian) in North Yorkshire and Cleveland. *Proceedings of the Yorkshire Geological Society*, **41**, 325–346.
Wright, J.K., 1978. *Stratigraphical and palaeoenvironmental studies of the Callovian rocks in North Yorkshire and Cleveland*. Unpublished PhD Thesis, University of London.
Wright, J.K. 1983. The Lower Oxfordian (Upper Jurassic) of North Yorkshire. *Proceedings of the Yorkshire Geological Society*, **44**, 249–281.
Wright, J.K. 1992. The depositional history of the Hackness Coral-Sponge Bed and its associated sediments within the Passage Beds Member of the Coralline Oolite Formation (Corallian Group; Oxfordian) of North Yorkshire. *Proceedings of the Yorkshire Geological Society*, **49**, 155–168.

4 References

Wright, J.K. 1996. Perisphinctid ammonites of the Upper Calcareous Grit (Upper Oxfordian) of North Yorkshire. *Palaeontology*, **39**, 433–469.

Wright, J.K. 2009. The geology of the Corallian ridge (Upper Jurassic) between Gilling East and North Grimston, Howardian Hills, North Yorkshire. *Proceedings of the Yorkshire Geological Society*, **57**, 193–216.

Wright, J.K., Bassett-Butt, L. & Collinson, M. 2014. Fatally bitten ammonites from the Lower Calcareous Grit Formation (Upper Jurassic) of NE Yorkshire, UK. *Proceedings of the Yorkshire Geological Society*, **60**, 1–8.

Wright, J.K. & Rawson, P.F. 2014. The development of the Betton Farm Coral Bed within the Malton Oolite Member (Upper Jurassic, Middle Oxfordian) of the Scarborough District, North Yorkshire, UK. *Proceedings of the Yorkshire Geological Society*, **60**, 123–134.

Young, G. 1817. *A History of Whitby*. Clark & Medd, Whitby, viii + 956 pp.

Young, G. & Bird, J. 1822. *A Geological Survey of the Yorkshire Coast*. Clark, Whitby, iv + 322 pp. (2nd edit. 1828).

Index

A

Aalenian 11, 13, 65
Abbotsbury Cornbrash Formation 12, 93, 95, 96, 98, 101, 104, 106
Africa 10
Albert Iron and Cement Works 42
Albian 4, 16, 124
Alpine movements 5
Alpine orogeny 4, 95
Alum Shale Member 10, 20, 36, 42, 44, 46, 48, 61, 62, 63, 65
Ampthill Clay Formation 4, 12, 15
Anglian 18
Aptian 124
Avicula Seam 33, 86

B

Bajocian 11, 65
Banded Shales 57
Barmston 24, 25, 151, 152, 154
Barmston-1 24, 30
Barremian 123, 124
Barton Marls 127, 129
Basement Till 18, 139, 145, 146, 147, 156, 157, 158
Bathonian 11, 12
Beacon Hill 131, 142
Beacon Hill Farm Marls 1–4 142
Beacon Hill Marl 131
Bempton Cliffs 16
Bempton zone 149
Berriasian 165
Berry Member 101
Betton Farm 107, 109, 111
Betton Farm Coral Bed 107
Betton Farm North Quarry 109
Betton Farm Quarries SSSI 107
Betton Farm South Quarry 107, 109, 111
Birdsall 111
Birdsall Calcareous Grit Member 14, 109, 111, 114
Bituminous Shales 10, 40, 42, 48, 49, 51, 58, 59, 61, 62
Black Band 17, 124, 125
Black Cliff 123
Blatterton 123
Blea Wyke Sandstone 7, 62, 65
Blea Wyke Sandstone Formation 62, 65
Bogmire Gill Member 73, 91, 92, 93
Boulby 20, 21
Boulby potash mine 31

Brackenberry Wyke 31, 33, 35, 36, 37
Bridlington 4, 18, 23, 30, 139, 148, 151, 158
Bridlington Bay 136, 140, 145
Bridlington Crag 139, 158
Bronze Age 20, 144, 148
Buckton 16
Bunter 24
Burnham Chalk Formation 16, 126, 130, 132, 134, 148
Burniston Bay 75, 76, 77, 78
Burniston Footprint Bed 77
Burniston Steps 76, 78

C

Calcareous Shale Member 7, 54, 56
Callovian 4, 12, 13, 95, 106
Campanian 145
Cannon Ball Doggers 38
Carboniferous 4, 21, 24, 28, 151, 156
Carr Naze 116
Castle Hill 84, 96
Cayton Bay 26, 29, 65, 88, 94, 95, 96, 101, 102, 104, 106
Cayton Bay Fault 95
Cayton Bay Waterworks 101, 106
Cayton Clay Formation 12, 96, 98, 101, 103
Cayton Cliff landslide 96
Cayton Fault 29
Cement Shales 46, 47, 62, 65
Cenomanian 17, 124, 164
Chalk 4, 15, 16, 17, 19, 123, 124, 126, 128, 129, 130, 132, 133, 134, 135, 136, 138, 139, 140, 144, 146, 148, 149, 150, 156, 157
Chalk Group 16, 124
Cleveland Anticline 5, 28, 65
Cleveland Basin 1, 4, 5, 6, 7, 10, 11, 12, 15, 17, 73, 90, 111, 126
Cleveland Dyke 17, 20, 82, 84, 85
Cleveland Hills 20, 33
Cleveland inversion 24
Cleveland Ironstone Formation 7, 20, 31, 33, 34, 40, 42, 43, 85
Cloughton 11, 24, 67, 68
Cloughton Formation 11, 20, 44, 65, 67
Cloughton Wyke 67, 69, 70, 71, 72
Cloughton Wyke Plant Bed 70
Common Hole 134
Commune Subzone 61
Coprolite Bed 121
Corallian Group 12, 20, 107
Coralline Oolite Formation 12, 13, 107, 109, 111

171

Index

Coral Rag Member 15, 107
Cordatum Zone 114
Cornelian Bay 87, 93, 94, 95
Cowbar Nab 31
Cowlam Hole 79
Coxwold-Gilling 149
Crassum Zone 62
Craven Fault Belt 4
Cromer Point 75, 78, 79
Crowe's Shoot Member 124
Curling Stones 38, 40

D

Danes Dyke Lower Marls 143
Danes Dyke Member 141, 143, 145
Danes Dyke Upper Marl 1 145
Daneswood Lower Marl 145
Daneswood Middle Marl 145
Daneswood Upper Marl 145
Davoei Zone 31
Deepdale Flint 130
Deepdale Lower and Upper Marls 130
Devensian 18, 19, 52, 72, 82, 109, 139, 145, 146, 147, 151, 153, 155, 157, 158, 159
Dimlington 139, 146, 156, 157, 158, 159
Dimlington High Land 156, 158, 159
Dimlington Silts 147, 156, 157, 158
Dimlington Stadial 18, 115, 151, 157
Discoscaphites binodosus Subzone 146
Dogger Formation 6, 10, 11, 21, 44, 46, 47, 48, 62, 65
Driffield 4, 148, 151, 156
Duckscar Quarry 82
Dulcey Dock 124

E

Easington Gas Terminal 156
East Ayton 107
East Cliff 44, 45, 46, 47
East Midlands Shelf 1, 4, 12, 15, 124, 126
East Nook 139, 140
East Nook Marl 1 140
East Pier, Whitby 44
East Scar 131
Ebberston 21
Edinburgh 1
Egton Bridge 82, 83
Eller Beck Formation 11, 44, 82, 95
Eskdale 21, 82, 83, 162
Eskdale Dome 83
Exaratum Subzone 10

F

Ferriby 16, 123, 124
Ferriby Chalk Formation 16, 124
Ferruginous Flint 127, 129, 131
Fibulatum Subzone 61
Filey 15, 149
Filey Brigg 19, 107, 109, 111, 112, 113, 114, 115, 116
Flamborough 16, 19, 22, 23, 126, 127, 132, 139, 143, 147, 148
Flamborough Chalk Formation 17, 132, 136, 138, 139, 144, 148
Flamborough Fault Zone 15, 22, 24, 30
Flamborough Head 16, 24, 30, 126, 132, 134, 135, 139, 140, 151
Flamborough Head Heritage Coast 3
Flamborough Moraine 149
Flat Scars 56
Fleet Member 12, 93, 101, 103
Flixton 149, 150, 166
Forge Valley 149, 150
Fox Cliff Siltstone Member 10, 62, 65
Foxholes 148

G

Ganton 150
Goathland 82, 83, 84
Great Thornwick Bay 126, 127, 130, 131
Grey Band 124
Grey Sandstone 10
Grey Shale Member 7, 10, 36
Gristhorpe Bay 103, 104, 105, 106
Gristhorpe Member 11, 67, 69, 71, 72, 96, 98, 99, 105
Gristhorpe Plant Bed 104, 105
Grobkreide 145
Grosmont 20, 83, 84, 85, 86

H

Hackness Rock Member 12, 95, 98, 101, 103, 104, 106
Hagenowia rostrata Zone 132, 137, 140
Hambleton Oolite Member 14, 109, 111, 114, 115
Hard Shales 46, 48
Hartendale Gutter 142, 143
Hartendale Marl 142
Hauterivian 121
Hayburn Wyke 65
Helwath Beck Member 70, 72, 103, 104, 105
Hertford River 150

Index

Hessle 19, 146
Hettangian 8, 54
High Stacks 134, 135, 136, 138
High Stacks Flint 134
Holderness 1, 17, 18, 19, 20, 117, 139, 140, 145, 146, 151, 152, 153, 154, 155, 156, 157
Holmpton 156
Hornsea 151, 154, 155
Hornsea Mere 152
Horse Shoe Rocks 105
Howardian-Flamborough Fault Belt 4, 5, 16, 126, 132, 139
Howardian Hills 4
Hull 20
Humber 23, 146, 156
Hundale Point 72, 73
Hundale Sandstone Member 72, 73, 101, 104
Hundale Scar 72
Hundale Shale Member 72, 73, 104
Hunmanby 24
Hunstanton Formation 15, 123, 124, 149
Hutton Buscel kame terrace 149

I

Ipswichian 36, 140, 146
Ironstone Shale Member 7, 54, 57, 58

J

Jamesoni Zone 58
Jet Rock 10, 36, 38, 40, 49, 58, 163
Jet Wyke 33, 35
Julian Park 84
Jump Down Bight 48

K

Kellaways Formation 12
Kellaways Sand Member 12
Kettle Ness 41, 43
Kettleness Member 33, 35, 37
Kettleness Sand 43
Khyber Pass 44
Kimmeridge Clay Formation 4, 15, 20, 118, 121, 149
Kimmeridgian 15
Kindle Scar 135
Kirby Misperton 21
Kirby Underdale 4
Kirkham Abbey carbonate 28
Kirkham Abbey Formation 21
Kirkham Gap 149
Kirmington 156
Knapton 21

L

Lake District 19
Lake Eskdale 82
Lake Pickering 18, 82, 149, 150
Lake Wheeldale 82, 84
Lambfold Hill Grit Member 93
Langdale Member 12, 95, 101, 103, 106
Langtoft 148
Langtoft Fault 144, 148
Laramide 4
Larvikite 158
Late Cimmerian movements 4
Lebberston Member 11, 67, 95, 96, 98, 105
Lias Group 7
Lincolnshire 19, 146, 156
Lisset 152
Little Thornwick Bay 127, 129
Long Bight 46, 47, 48
Longhorne Wyke 78
Long Nab 77
Long Nab Member 11, 12, 75, 76, 77, 78, 81, 92, 93, 94
Longwood Marl 146
Lower Calcareous Grit Formation 12, 13, 96, 100, 101, 102, 103, 109, 111
Ludborough Flint 131

M

Maidlands Lower Marls 143
Maidlands Upper Marl 1 143
Main Alum Shales 46, 48, 65
Main Seam 35, 37, 85
Malton 12, 14, 111
Malton Oolite Member 15, 107
Mappleton 154, 156
Margaritatus Zone 31, 62
Market Weighton 12, 15
Market Weighton High 4, 12, 13, 14, 15, 111
Marsupites testudinarius Zone 140, 143, 145
Melton Ross Marl 127, 129, 130
Mesolithic 150, 153
Middle Calcareous Grit Member 14
Middle Cliff 118, 122, 123
Middle Drab 155
Middle Estuarine Series 44
Mid North Sea High 5, 11, 90
Millepore Bed 11, 67, 95, 96, 98, 99, 103, 105
Millers Nab 56, 57

Index

Millstones 38, 40, 42, 49
Miocene 4, 5
Molk Hole 134
Moor Grit Member 11, 67, 73, 74, 75, 81, 90, 91, 92, 93, 94, 103, 104, 105
Mulgrave Shale Member 10, 36, 40, 42, 48, 49, 58, 59, 61, 82
Mull 17
Murk Esk 85, 86

N

Netherlands 10
New Closes Cliff 118
Newtondale 11, 82, 149
Norfolk 16, 19
North Germany 10, 123
North Landing 126, 128, 130
North Sea 1, 11, 15, 18, 21, 22, 65, 121, 123, 134
North Sea Basin 7, 15, 121
North Sea Lobe 18, 140, 144, 145, 149, 155, 158
North York Moors 2, 17, 18, 20, 21, 31, 40, 44, 52, 67, 75, 82
North York Moors National Park 2
North Yorkshire & Cleveland Heritage Coast 2
Norway 151

O

Oceanic Anoxic Event (OAE) 10, 17, 48, 124
Old Nab 33, 35, 37
Old Quay Rocks 114, 116, 117
Osgodby 98
Osgodby Fault 29, 96, 101
Osgodby Formation 12, 95, 96, 100, 101, 102, 103, 105
Osgodby Hill 96, 106
Osgodby Point 11, 65, 67, 94, 95, 96, 98, 99
Out Newton 156
Ovatum Band 42, 48, 49, 59, 61
Oxford Clay Formation 12, 95, 96, 98, 100, 101, 102, 103, 105, 106
Oxfordian 4, 12, 14, 15, 95, 114, 115
Oxynotum Zone 57

P

Palaeolake Flixton 149
Pallasioides Zone 15
Paris Basin 10
Peak 52, 54, 61, 62, 63, 64
Peak Alum Works 66

Peak Fault 5, 10, 26, 52, 58, 59, 60, 62, 64, 65, 70
Peak Mudstone Member 10, 62, 65
Peak–Red Cliff Fault Zone 96
Peak Steel 58, 59, 65
Peak Stones 59
Peak Trough 5, 24, 26, 27, 29, 64, 65, 75, 96, 103, 104
Pecten Seam 35, 85, 86
Pectinatus Zone 15
Pennine High 5
Pennine landmass 7
Penny Nab 31, 33
Penny Nab Member 33
Permo-Triassic 151, 155
Peterborough Member 12
Pickering system 24
Pleistocene 17, 52, 112, 138, 139, 140, 144, 146, 156, 157, 158
Pliensbachian 7, 54, 57, 62
Pliensbachian Global Stratotype Section and Point 57
Port Mulgrave 31, 36, 38, 39, 40
Posidonienschiefer 10
Pyritous Shale Member 7, 54, 57

Q

Quaternary 1, 18, 22, 107, 136, 138, 144
Queen Rock 124

R

Rail Hole Bight 48
Raisdale Seam 33, 35
Raricostatum Zone 57
Ravendale Flint 130, 131
Ravenscar 10, 20, 26, 28, 29, 52, 54, 60, 62, 65, 70
Ravenscar brickworks 64
Ravenscar Group 11, 20, 44, 65, 67, 69, 75, 96, 100, 103
Ravenscar Shale Member 73, 74, 93, 104
Redcar Mudstone Formation 7, 24, 52, 54, 55, 58, 62
Red Chalk 15, 124
Red Cliff 65, 94, 101, 102, 103, 106
Red Cliff Fault 5, 24, 29, 94, 96, 101, 102, 103, 104
Red Cliff Hole 124
Red Cliff Hole Member 124
Red Cliff Rock Member 12, 95, 96, 98, 101, 103
Redscar Hole 42, 43

Index

Reighton 3, 118, 124
Rhaetian 4
Rhaetic 24
Rhomb Porphyry 158
River Humber 156
Robin Hood's Bay 7, 19, 24, 26, 27, 28, 52, 54, 55, 56, 57, 58, 62, 64, 66
Robin Hood's Bay Dome 5, 24, 28, 62, 64
Robin Lythes Hole 131
Rødby Formation 124
Roos 159
Rosedale 21
Rosedale Wyke 32, 38
Rotunda Museum 87
Rowe Formation 17
Rudston 148
Rudston churchyard 148
Rudston monolith 148
Runswick Bay 40, 41, 42, 43, 49
Ryazanian ii, 16, 164

S

Saintoft Member 109, 111
Saltburn 3
Salt Pans 67, 71, 74
Saltwick Bay 44, 45, 48, 49, 50, 51
Saltwick Formation 11, 20, 36, 42, 44, 46, 47, 48, 62, 65
Saltwick Nab 44, 48, 49, 50
Santonian 132, 145
Scalby Bay 12, 75, 76, 78, 81
Scalby Lodge 79
Scalby Mills 3
Scalby Ness 69, 81
Scalby Plant Bed 79
Scandinavia 19
Scarborough 1, 12, 14, 15, 20, 26, 65, 67, 73, 74, 87, 89, 90, 91, 93, 95, 103, 107, 149
Scarborough Formation 11, 67, 70, 72, 73, 91, 92, 93, 96, 101, 102, 103, 104, 105
Scarburgense Subzone 95
Schistes Carton 10
Scotland 17, 19, 151
Seamer delta 149
Seamer-Scarborough Valley 149
Seaton Drift 38
Selwicks Bay 16, 132, 133, 134, 135
Sewerby 3, 18, 19, 136, 138, 140, 144, 145, 147, 157
Sewerby Buried Cliff 146, 157
Sewerby Member 145, 146

Sewerby Raised Beach 146
Sewerby Steps 136, 138, 140, 144, 146
Shap Fell 52
Shap Granite 52, 126
Sheep Stones 36
Sherwood Sandstone 24
Silhowe 84, 85
Siliceous Shale Member 7, 54, 56, 57
Sinemurian 54
Skipsea 152, 153, 154
Skipsea Till 19, 117, 118, 126, 136, 139, 140, 141, 145, 146, 151, 152, 153, 154, 155, 156, 157, 158, 159
South Bay, Scarborough 87, 91, 103
Southern North Sea Basin 7
South Landing 136, 137, 139, 140, 141, 142, 144, 145, 147, 151
South Landing Member 137, 139, 141, 142
Speeton 16, 119, 121, 124, 125
Speeton Beck 118, 123, 124
Speeton Clay Formation 15, 20, 118, 119, 120, 121, 123, 124, 149
Speeton Cliffs 123, 124
Sphenoceramus lingua Zone 140, 142, 145, 146
Spindle Thorne Limestone 72, 73, 101, 104
Staithes 7, 20, 31, 32, 33, 34, 36, 38, 43, 58, 62, 85
Staithes Beck 31
Staithes Harbour 33
Staithes Sandstone Formation 7, 31, 33, 34, 58, 62
Staple Nook 16, 149
Starr Carr 150
Staxton 148, 149
Staxton Hill 148, 149
Sternotaxis plana Zone 130, 131
Stewartby Member 12
Stone Cross 84
Stoup Beck 54
Sycarham Member 11, 95, 105

T

Tabular Hills 18, 149, 166
Tenants' Cliff 100, 103
Tenants' Cliff Member 101
Terebratulina lata Zone 126, 130
The Combe 62
Thorndale Shaft 36
Thornwick Bay 126, 127, 128, 129, 130, 131
Thornwick Nab 127, 129
Top Jet Dogger 38

Index

Topman Steel 40, 42, 43
Triassic 4, 5, 7, 24, 65, 151, 155
Triple Tabular Flints 130, 131
Tunstall 156
Turneri Zone 56
Two Foot Seam 85

U

Uintacrinus anglicus Zone 145
Uintacrinus socialis Zone 140, 142, 143
Ulceby Marl 130, 131
Ulceby Oyster Bed 131
Ulrome 152
Upgang 18
Upper Calcareous Grit Formation 12, 15
Upper Drab Till 155
Upper Pseudovertebrae 38, 40

V

Valanginian 121
Vale of Pickering 15, 18, 20, 21, 24, 108, 148, 149, 150
Vale of York 18, 149
Variscan Unconformity 24, 28
Vertebrale Subzone 115
Volgian 4, 14, 15

W

Wealden facies 15
Weather Castle Member 124, 125
Welton Chalk Formation 16, 17, 124, 126
West Cliff 44, 132, 133, 134, 135
West Nook 139, 140, 141, 142
West Nook Marls 142
Weymouth Member 12, 95
Whalestones 38, 42
Wheeldale 82, 84
Whitby 1, 5, 10, 18, 21, 22, 23, 24, 25, 39, 40, 44, 45, 46, 47, 49, 50, 83
Whitby Abbey 20
Whitby Fault 44
Whitby Mudstone Formation 7, 10, 31, 36, 40, 44, 52, 58, 64, 82, 85
Whitby Museum 48
Whitby Plant Bed 47
White Nab 93
White Nab Ironstone Member 73, 91, 92, 93, 103, 104
Wiltshire 12
Wine Haven 57, 58, 165
Withernsea 156, 157
Withernsea Till 19, 117, 154, 156, 157, 158, 159
Withow Gap 152, 153, 154
Wolds 15, 16, 17, 18, 20, 139, 146, 148
Wolstonian 147, 157
Wool Dale 116, 117
Wrack Hills 42
Wykeham lakes 149
Wykeham Moraine 149

Y

Yedmandale Member 13, 104, 111
Yellow Sandstone 10, 62
Yons Nab 11, 67, 89, 99, 101, 103, 104, 105, 106
Yons Nab Beds 11, 67, 69, 98, 99, 103, 105

Z

Z2 Formation 24
Zechstein 5, 24, 27, 28, 30, 54
Zechstein Group 24

176

Other Guides published by the Geologists' Association:

No.	
2	Geology of the Lake District
6	University Towns - Liverpool
7	Geology of the Manchester Area
10	North Cornwall - Bude to Tintagel
22	Dorset Coast
26	Geology of the Peak District
27	Geology of South Shropshire
31	West Cornwall
32	Isle of Arran
38	The Plymouth Area
40	Anglesey
41	Jersey
43	Costa Blanca, Spain
45	Onny Valley, Shropshire, Geology Teaching Trail
46	The Isle of Man
47	Coastal Landforms of West Dorset
49	Tenerife
50	Southern Cyprus
51	Island of Bute, Scotland
52	Iceland
53	Eastern and Central Jamaica
54	The Aberystwyth District
55	Early Cretaceous Environments of the Weald
56	The Castleton Area, Derbyshire
57	The Chalk of Sussex and Kent
59	Hadrian's Wall
60	Isle of Wight
61	Geology of the Western Front
62	Geology of Lanzarote
63	Geology of the East Midlands
64	D-Day Landings, Normandy, 1944
65	The Origins of Stratigraphy
66	Watchet, Somerset
67	The Dalradian of Scotland
68	Geology of London
69	The Roadside Geology of Wales
70	Barcelona: urban excursion guide
71	Bristol Region: Quaternary Geology and Geomorphology
72	Devonshire Marbles 1. Understanding the marbles 2. Their geology, history and uses
73	South Devon Coast

Rockwatch Guide No. 1 'A Pocket Guide to the London Clay exposed on the North Shore of the Isle of Sheppey, Kent'

Additional guides and current prices are available from: **The Executive Secretary, Geologists' Association, Burlington House, Piccadilly, London W1J 0DU**
Telephone: 020 7434 9298
E-mail: admin@geologistsassociation.org.uk
Website: www.geologistsassociation.org.uk